Cross-Sectional Anatomy for Computed Tomography

Michael L. Farkas

Cross-Sectional Anatomy for Computed Tomography

A Self-Study Guide with Selected Sections
from Head, Neck, Thorax, Abdomen, and Pelvis

With Collaboration from S. Kubik
Translated by James D. Fix
Foreword by Elliot K. Fishman

With 93 Figures

Springer-Verlag New York Berlin Heidelberg
London Paris Tokyo

Michael L. Farkas, M.D.
Diplomate, American Board of Plastic Surgery, 3013 Bern, Switzerland

James D. Fix, Ph.D.
Professor, Department of Anatomy, Marshall University, School of Medicine, Huntington, West Virginia 25704, USA

Stefan Kubik, M.D.
Associate Professor, Head, Department of Macroscopy, Institute of Anatomy, University of Zurich-Irchel, CH-8057 Zurich, Switzerland

On the front cover: Fig. 30. Planes of section through thorax, p. 43

Library of Congress Cataloging-in-Publication Data
Farkas, M. (Michael)
   Cross-sectional anatomy for computed tomography.
   Translation of: Querschnittanatomie zur Computer-
tomographie.
   Bibliography: p.
   1. Anatomy, Human—Atlases. 2. Tomography—Atlases.
I. Kubik, Stefan. II. Title. [DNLM: 1. Anatomy—
atlases. 2. Anatomy—programmed instruction.
3. Tomography, X-Ray Computed—atlases. 4. Tomography,
X-Ray Computed—programmed instruction. QS 18 F229q]
WM25.F3713 1988   611'.0022'2   88-6554

Translation of *Querschnittanatomie zur Computertomographie*/Eine Einführung mit ausgewählten Schnitten aus dem Kopf-, Hals-, Brust- und Beckenbereich. © Springer-Verlag Berlin Heidelberg, 1986.

Typeset by Bi-Comp, Inc., York, Pennsylvania.
Printed and bound by Arcata Graphics/Halliday Lithograph, West Hanover, Massachusetts.

9 8 7 6 5 4 3 2 1

ISBN-13: 978-1-4613-8784-8   e-ISBN-13: 978-1-4613-8782-4
DOI: 10.1007/ 978-1-4613-8782-4

Foreword

The clinical acceptance of computed tomography (CT) as an integral part of our diagnostic armamentarium was based on its ability to display cross-sectional anatomy with near anatomic precision. However, the radiologist must first be knowledgeable of the complexities of normal anatomy before he can truly make full use of this technology.

Michael Farkas has truly made our task as radiologists easier. As noted in the preface, the book carefully correlates representative CT slices with corresponding anatomic cross-sections. Schematic line drawings are also generously used to illustrate particularly complex anatomic regions and help the reader obtain a correct perspective on these more difficult regions. The book successfully presents a clear perspective on the anatomy we see daily in using cross-sectional imaging techniques. This book will prove useful as a learning guide for the uninitiated, and as a reference for the more experienced. Either way, it is an important contribution to our literature.

Elliot K. Fishman, M.D.
Associate Professor
Director, Division of
   Computed Body Tomography
Department of Radiology and
   Radiological Science
The Johns Hopkins Medical Institutions
Baltimore, Maryland

Preface

This self-study guide is intended to explain the basic principles of computed tomography (CT) and to stress the importance of cross-sectional anatomy, using cardinal transverse sections of the body to interpret CT-images.

The objectives of this manual are to enable the reader:

1. To understand the basic principles of CT.
2. To study the cross-sectional anatomy of selected cardinal levels of the head, neck, and trunk.
3. To learn the orientation and interpretation of CT-images with the aid of corresponding cross-sectional preparations.

At most medical schools, students do not receive enough instruction in cross-sectional anatomy because of curricular time constraints.

With the advent of CT the importance of cross-sectional anatomy cannot be overestimated. Cross-sectional anatomy is the foundation for understanding and interpreting the CT-image. Most textbooks of cross-sectional anatomy and CT are written for the clinical specialist, contain too much detail, and hence cannot be used for self-study. This work provides the reader with a brief and easily understandable self-study guide to cross-sectional anatomy and CT.

A learning program in its present form was considered to be one of the best methods for self-study. In the beginning of this text the principles of CT are explained in a clear and simple way. An optimal number of cross sections were selected so that a representative series through the body regions could be achieved. Ten sections from the head, two from the neck, three from the thorax, three from the abdomen, and three from the pelvis were used. Sections from both the male and the female pelvis are presented. The relative number of CT scans included for the head/neck, chest, and abdominal regions reflect the complexity of the structures contained in these areas, and not the relative emphasis in clinical applications of CT.

Line drawings were used with the sections to aid the reader in his orientation and three-dimensional visualization. After selection of the transverse sections, corresponding CT-scans were chosen to match the anatomical sections.

From the CT-scans only those structures were labeled that could be positively identified. Since this manual is intended to serve as a self-instructional program it was decided to clearly label only the most important structures.

I cordially thank Professor S. Kubik for his support in the composition and compilation of this text as well as for his photography of the cross sections. I thank Professor A. Valvanis for his aid and advice in the selection of the CT-pictures. I thank Ms. M. Müller for her assistance in the photography laboratory, Mr. A. Lange for his help in preparing the anatomical sections, and Dr. B. Szarvas for his aid in labeling the pictures.

Bern, Switzerland

Michael L. Farkas

Contents

1 Introduction

Since the introduction of computed tomography (CT) to clinical medicine by Hounsfield et al. (1973) the importance of cross-sectional anatomy has significantly increased. The evaluation and interpretation of CT images are dependent upon a detailed knowledge of cross-sectional anatomy. A knowledge of cross-sectional anatomy coupled with an understanding of the principles of computed tomography form the foundation needed to evaluate the CT images.

It is the goal of this work to illustrate these two prerequisites. The main emphasis is placed on the study of cross-sectional anatomy. We realize that a series of cross-sectional anatomical preparations cannot match exactly with a series of CT images in the same plane of section. We have used, however, CT pictures from the collection of the Central Institute of Diagnostic Radiology of the University of Zürich which corresponded as closely as possible to the anatomical cross-sections.

We have used CT studies from live patients instead of making tomograms from cadavers, which would manifest artifacts of fixation. Blood vessels of cadavers contain fixative instead of blood; also, the lungs are filled with fluid as seen with edema.

In a tomogram of a cadaver section the air in the blood vessels, subarachnoid spaces, and serous spaces (pleural, and peritoneal spaces) would appear as unnatural artifacts.

In CT the blood vessels, secretory organs, and the gastrointestinal (GI) tract can be enhanced by the use of contrast media as in conventional radiographs.

Method of Preparation of Anatomical Cross-Sections

The anatomical cross-sections used in this text were taken from two female cadavers and one male cadaver. After embalming, the cadavers were frozen for 24h at −20°C in a deep freezer. The cross-sections were cut in the frozen condition with a band saw. Sections through the head were cut 1-cm thick, those through the rest of the body 2–3-cm thick. The thawed sections were then photographed. Since the CT image shows the plane of section from the caudal side, as if seen by the observer from the foot of the patient, the anatomical sections were correspondingly photographed from the caudal side.

CT images of the head are an exception and are presented to the observer as seen from the cephalic side.

2 Principles of Computed Tomography

2.1 Overview

Using conventional radiologic techniques it is not possible to produce an image of one plane of section. This is due to the projection mode, in which all planes of the irradiated body segment are projected on top of each other. A *tomogram* is a sectional roentgenogram, an image of one selected plane of section, and is made in the following manner. By giving the x-ray tube a curvilinear motion during exposure synchronous with the recording plate but in the opposite direction, the shadow of the selected plane remains stationary on the moving film while the shadows of all other planes have a relative displacement on the film and are therefore obliterated or blurred. Using this tomographic study with the patient in the recumbent position one obtains frontal sections. In *transverse tomography* the x-ray tube remains stationary and the seated patient and the film plate are rotated around the longitudinal axis of the patient. This type of tomography was replaced by computed tomography.

The CT scanner is an x-ray machine which generates sectional images of the body or cross-sections of organs with the aid of a computer and without x-ray film. For this reason the method is called computer assisted tomography (CT). In CT the x-ray beam which passes through the body is registered by detectors and not on film.

These detectors are capable of measuring the absorption coefficients of various tissues that differ from each other in a range of approximately 1%, with much higher resolution than that possible with x-ray film. Through rotation of the x-ray tube and the detector system, which is located opposite the x-ray tube, both in a plane perpendicular to the long axis of the body, one obtains a transverse (cross-sectional) body image. After exposure the computer subparcels each picture into a matrix of individual dots. For each matrix point (*pixel*) the computer calculates an individual absorption value. For each section there are approximately 10^5 data points, which are stored and evaluated by the computer. In the process of CT image reconstruction, matrix dots of varying gray tones corresponding to the different absorption values are assigned to the examined cross-section and displayed on a television (TV) monitor. Structures with high x-ray absorption appear light, those with low x-ray absorption appear dark.

Conventional *transverse* tomography is incapable of producing the sharp images that are seen in CT horizontal sections and so valuable in diagnostic radiology.

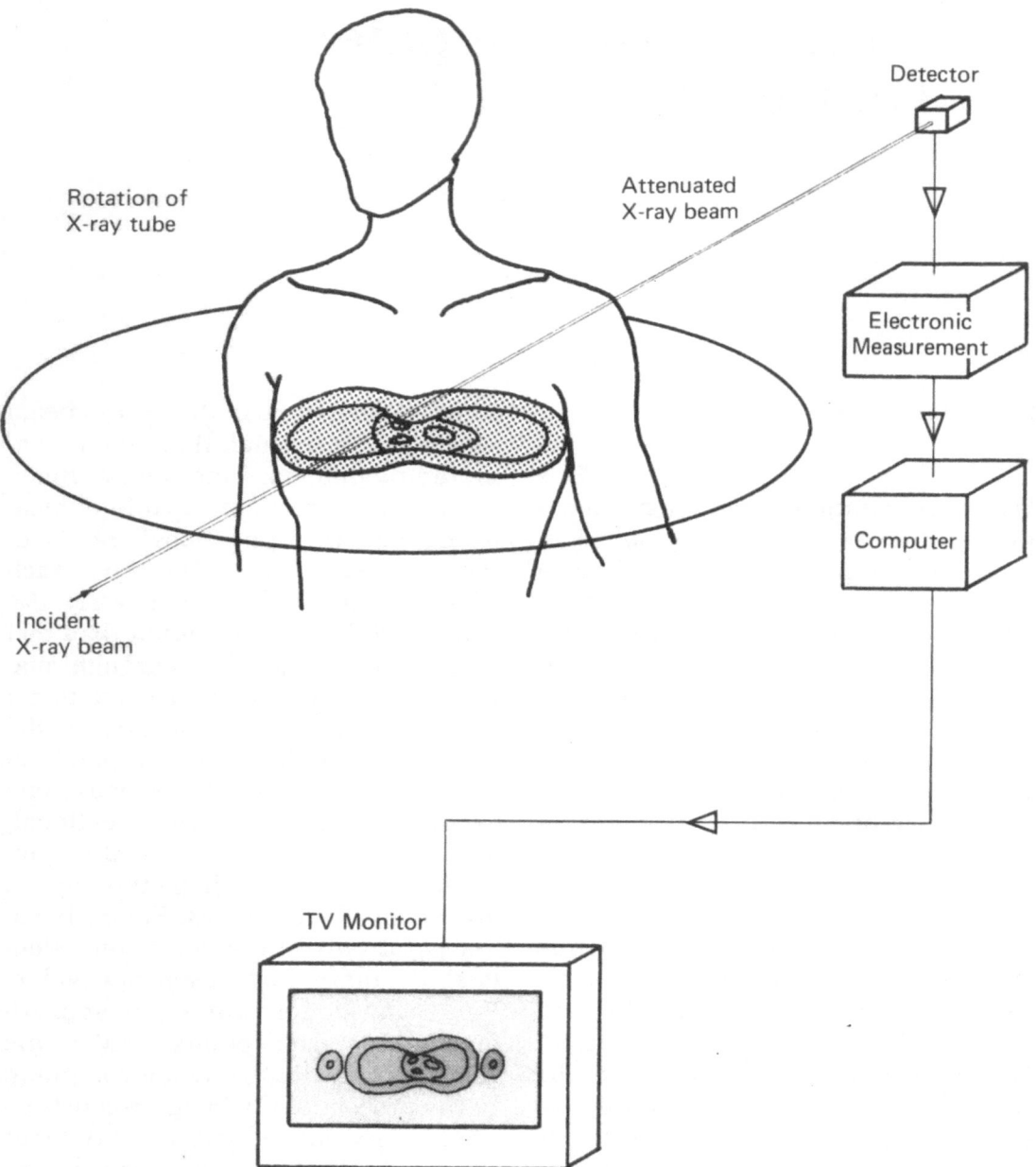

Fig. 1. Principle of computed tomography.

2.2 Data Acquisition

The principle of CT data acquisition can best be explained using as an example the cylinder (Figs. 2 and 3).

In scanning with CT the body (cylinder) is x-rayed layer by layer from various directions either parallel to or around the transverse axis. The attenuated x-ray beam is registered by detectors.

The x-ray beams are cone-shaped with a breadth of 1–3 mm and a height

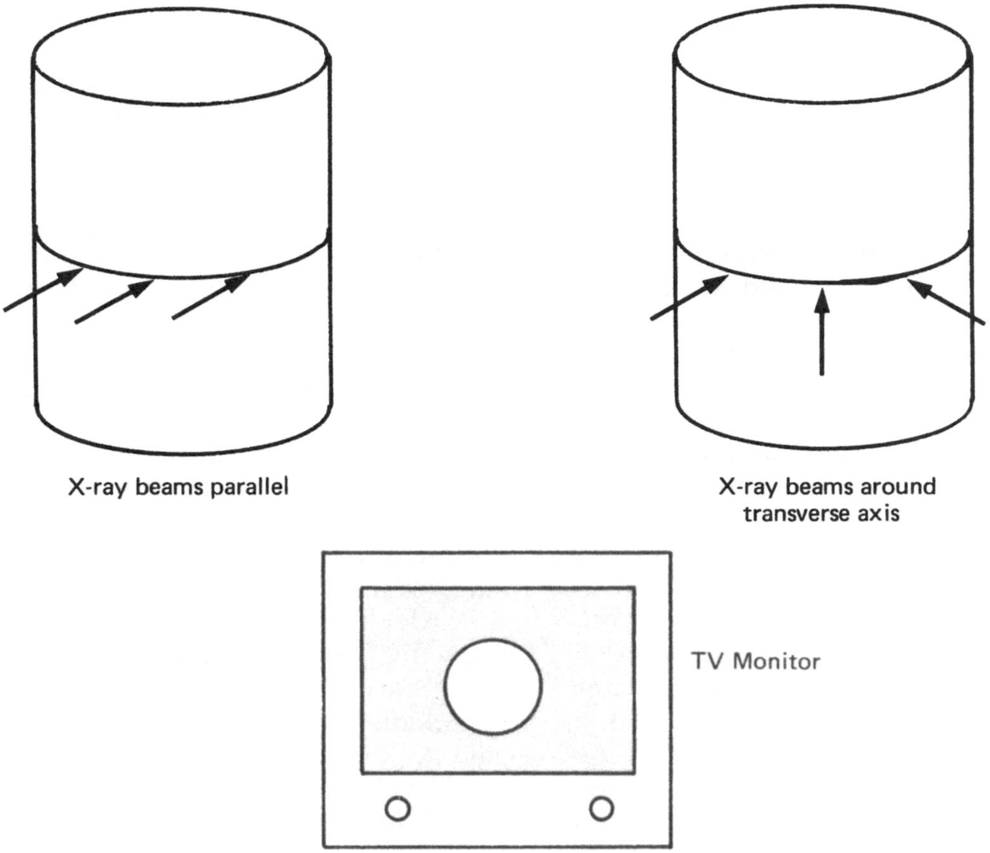

X-ray beams parallel

X-ray beams around
transverse axis

TV Monitor

Fig. 2. Principle of data acquisition.

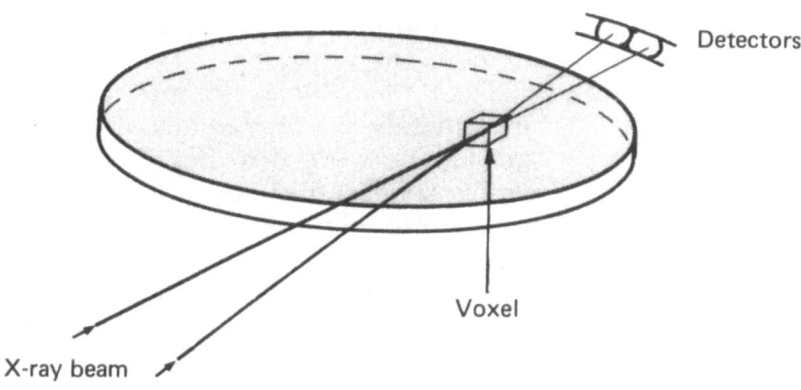

Detectors

Voxel

X-ray beam

Fig. 3. Slice of a cylinder.

of 2–10 mm. As x-rays pass through the body they are attenuated corresponding to the density of the tissue. For example the lung (containing air) attenuates the x-ray beam much less than does dense tissue such as muscle. Based on the degree of attenuation of the x-ray beam through various tissues, the tissues are assigned various absorption coefficients. The detectors in the CT machine are capable of registering minimal differences in intensity of the x-ray beams after they pass through tissue. Therefore it is possible even in the same tissue to differentiate areas that have a higher or lower concentration of water (different absorption coefficients), e.g., a localized focus of cerebral edema. Based on the registered absorption coefficients of the scanned slices of the cylinder, measured values can be assigned to the various unit volumes within the slice. Accordingly one may visualize this tissue slice of a cylinder as being made up of many (10^n) unit volumes. These unit volumes are called *voxels* (volume elements).

The slice of tissue, consisting of 10^n unit volumes, is systematically irradiated from different directions by 10^{xn} cone-shaped x-ray beams.

Utilizing this matrix system of irradiation it is possible to assign absorption coefficients to all of the volume elements.

2.3 CT Image Production

The CT image is presented in two dimensions. For this reason the individual volume elements are shown as image dots, which are provided with measurement values corresponding to the absorption coefficients. The individual image dots are called pixels. If one relegates the image dots (pixels to their locations, which are marked with measurement values, a mosaic is created which is composed of numbers. If the human eye is to reidentify the structure represented by the numbers, the numbers must again be transformed into easily identifiable structures.

As previously mentioned the individual pixels serve as a matrix for the CT image. Each pixel is assigned a specific absorption coefficient. Therefore the CT matrix reproduces the distribution of the absorption coefficients of the scanned layer.

In order to use a more convenient system than the absorption coefficients the arbitrary CT scale was introduced. The calculated CT values are proportional to the absorption coefficients. The CT values for water, air, and bone are as follows:

Water	0
Air	-1000
Bone	$+1000$.

All other tissues have CT values in the CT scale that corresponds to their absorption coefficients.

The CT values of all image dots are presented on the TV monitor as gray tones.

The human eye can differentiate approximately 30 gray tones. If all 30 gray tones were distributed over the entire spectrum of the CT scale, we could only differentiate those structures that differed from each other by a CT value in excess of 60. This would naturally be unsatisfactory. In order to achieve good contrast-resolution of structures one uses the technique of electronic windowing. In this method

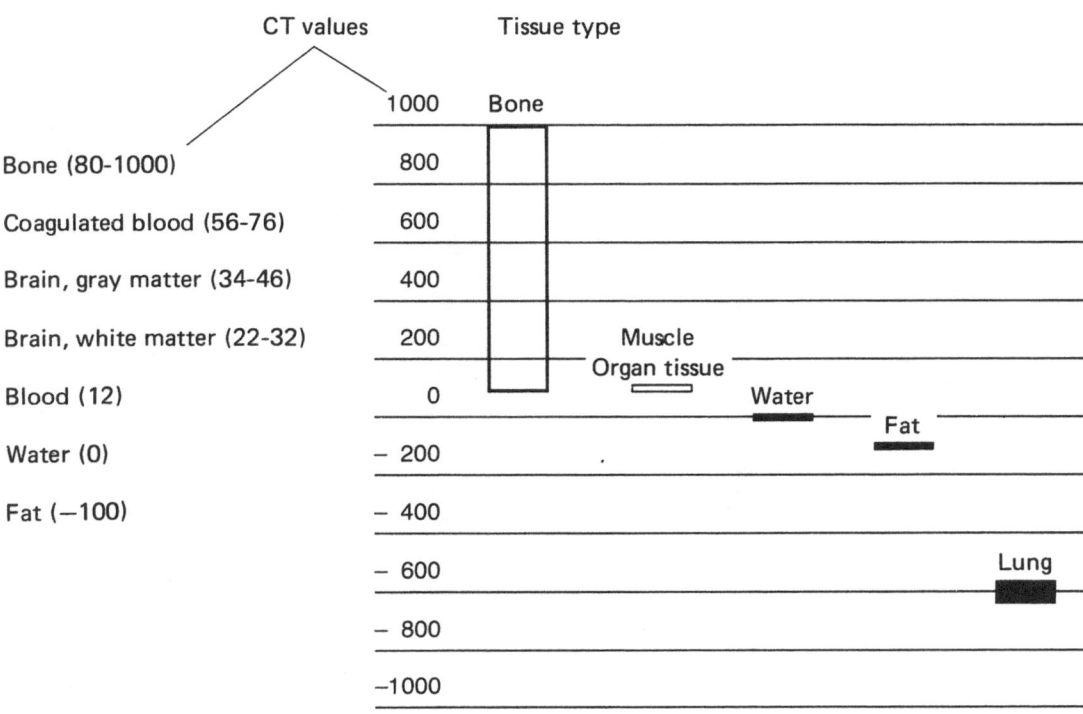

Fig. 4. Assignment of CT values to corresponding tissue types.

only a selected range of the CT scale in gray tones is used. Four CT values are allotted to each gray tone gradation. It is therefore possible to display a *range* (e.g., 30 × 4 = 120 CT values) in 30 different gray gradations. Everything outside the selected range of the scale appears white if it lies on the positive side, and black if it lies on the negative side. Figure 4 demonstrates CT value assignment to corresponding tissue types.

For example, if one wants to demonstrate lung tissue (Fig. 5) the "window level" control is set at −600, thus giving a range from −540 to −660. Moreover, this window width can be adjusted. The electronic windowing also makes possible a high contrast resolution for the human eye.

2.4 CT Systems

There are various types of CT machines or systems, which differ in the number and arrangement of detectors which lie opposite to the x-ray tube.

In the *single detector* system (Fig. 6) the x-ray tube and detector move in parallel during scanning.

There are two types of *multidetector* systems, utilizing either mobile detectors or stationary detectors. With the *mobile detector* system (Fig. 7) the x-ray tube and the detectors rotate parallel to each other during exposure.

With the *rotating fan beam* system (Fig. 8) the detector devices are stationary and the x-ray tube rotates around the axis of the patient.

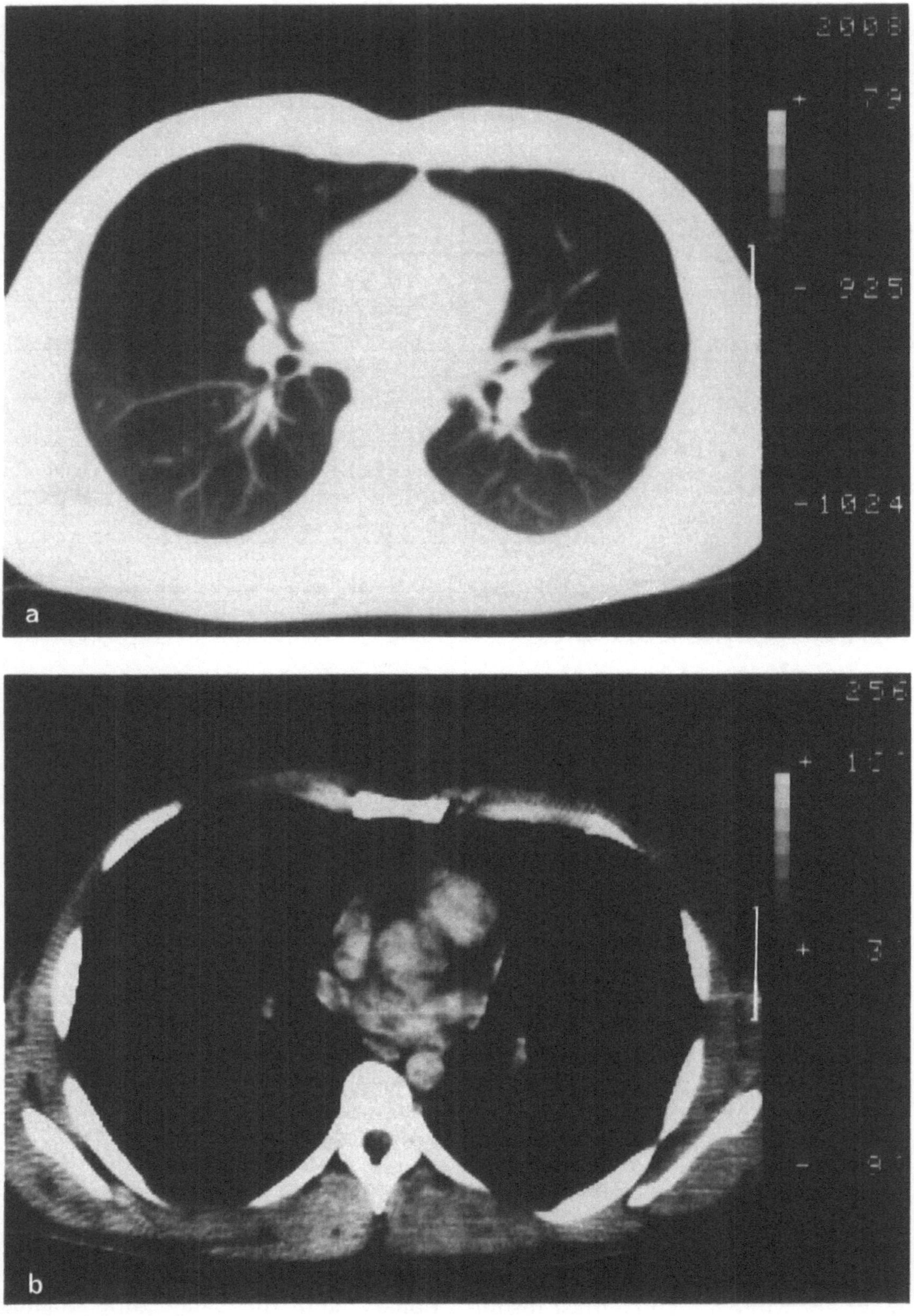

Fig. 5.*a*. Lung with lung window. *b*. Lung with mediastinal window.

X-ray tube   Detecto

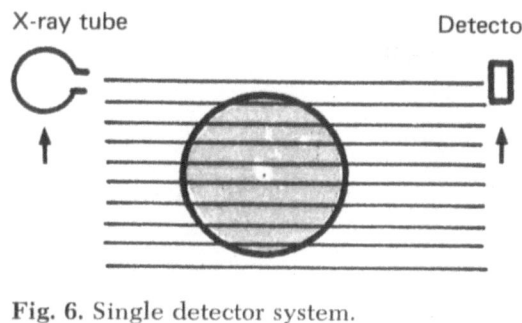

Fig. 6. Single detector system.

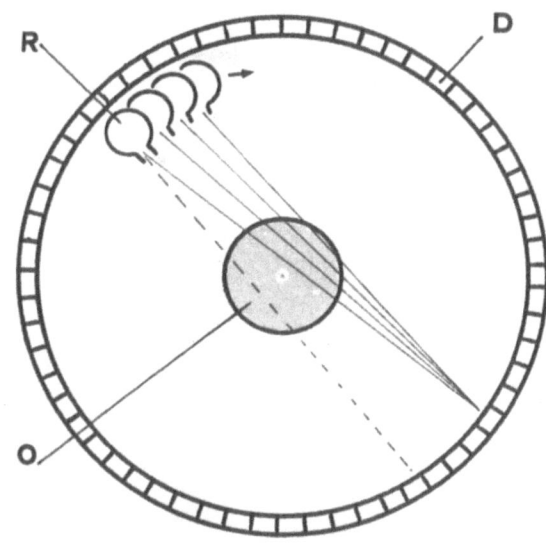

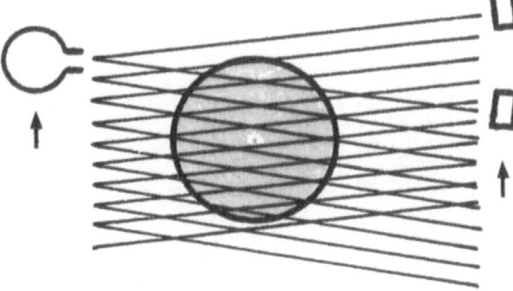

Fig. 7. Mobile multidetector system.

Fig. 8. Rotating fan beam system, *R*, x-ray tube; *D*, detector; *O*, subject.

3 Cross-Sectional Anatomy with Corresponding CT Images

3.1 Head, Brain, and Visceral Cranium

In CT "true" horizontal[1] sections are only used in scanning the neck and trunk regions of the body. In the head region the planes of section project obliquely. These oblique planes are offset from the infraorbitomeatal line[2] (a line tangent to the inferior orbital margin and the superior margin of the external acoustic meatus) by an angle of 30° (Fig. 9).

---

[1] In the United States sections cut parallel to Reid's base line are referred to as horizontal sections.

[2] In the Anglo-American literature the term *Reid's base line* is used. This line is tangent to the inferior orbital margin and the center of the external acoustic meatus.

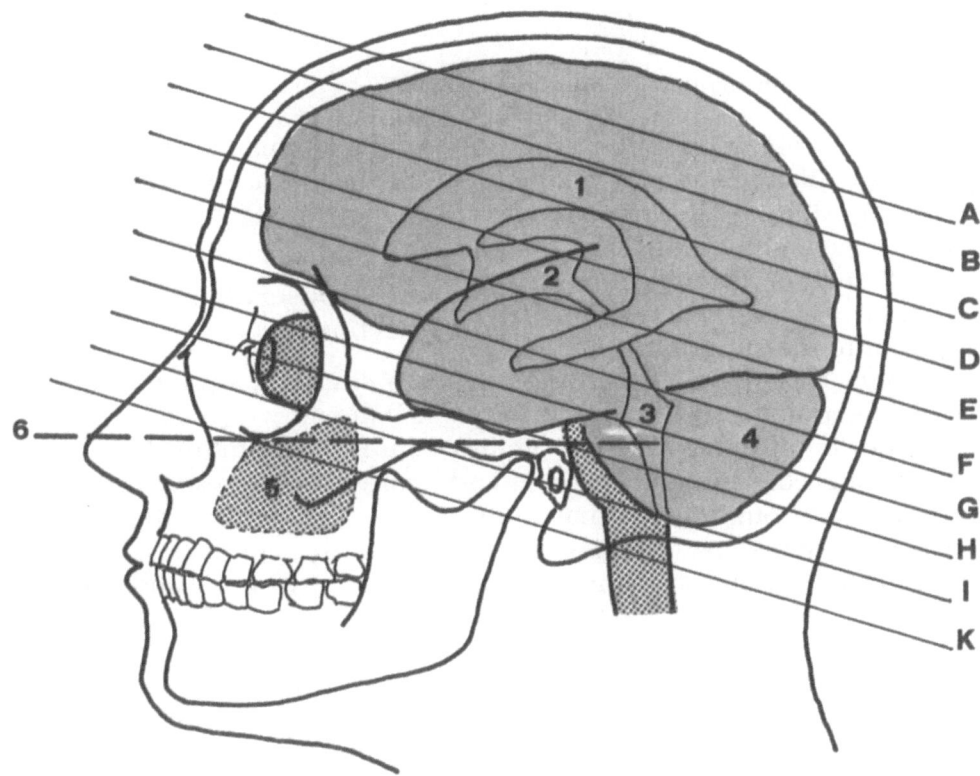

Fig. 9. Projection of parallel planes of section (*A–K*) on the lateral aspect of the head.

1. Lateral ventricle	*3.* Fourth ventricle	*5.* Maxillary sinus
2. Third ventricle	*4.* Cerebellum	*6.* Reid's base line

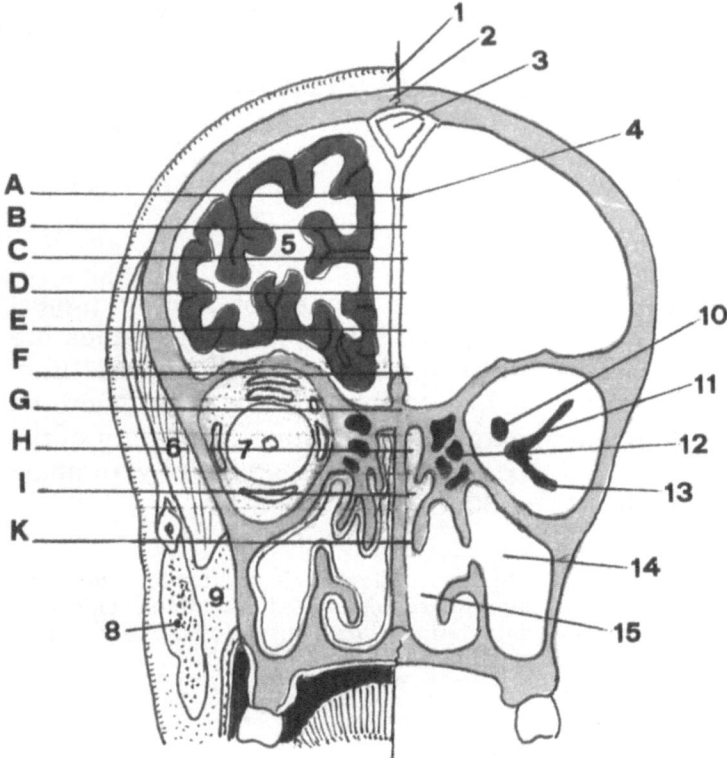

Fig. 10. Frontal section through the visceral cranium.

A–K Planes of section
1. Scalp
2. Skull cap
3. Superior sagittal sinus
4. Falx cerebri
5. Frontal lobe
6. Temporalis muscle
7. Bulbus oculi
8. Masseter muscle
9. Bichat's fat pad
10. Optic foramen
11. Superior orbital fissure
12. Ethmoid air cells
13. Inferior orbital fissure
14. Maxillary sinus
15. Nasal cavity

A rough orientation to the structures cut in this plane of section can be obtained by studying a schematic of the lateral aspect of the skull (Fig. 9) and a frontal section through the visceral cranium (Fig. 10). All sections (*A–F*) superior to the superior orbital margin pass only through the skull cap and brain. Sections which lie inferior to the superior orbital margin pass anteriorly through the visceral cranium and posteriorly through the cranial vault.

Sections of the visceral cranium pass through the nasal and paranasal cavities next to the midline, laterally, and in rostrocaudal extent first through the orbits and then through the maxillary sinuses (Fig. 10). Sections through the cranial vault pass through the petrous portion of the temporal bone, the external acoustic meatus of the tympanic portion of the temporal bone, the cerebellar hemispheres of the posterior cranial fossa, and the pons and me-

dulla oblongata which lie on the clivus of the basilar portion of the occipital bone (Fig. 9).

3.1.1 Observations on Supraorbital Sections

In studying supraorbital sections (*A–E*) the only changes noted are in structures of the brain (Fig. 11). The structure of the scalp and skull cap (calvaria) remain the same. In all sections through the calvaria the internal lamina, diploë, and external lamina can be seen. The falx cerebri is seen between the cerebral hemispheres, depending on the height of section, and the tentorium cerebelli is seen around the periphery of the cerebellum separating it from the occipital and temporal lobes (Fig. 9). The superior sagittal sinus is seen transected twice, anteriorly and posteriorly (Fig. 11a).

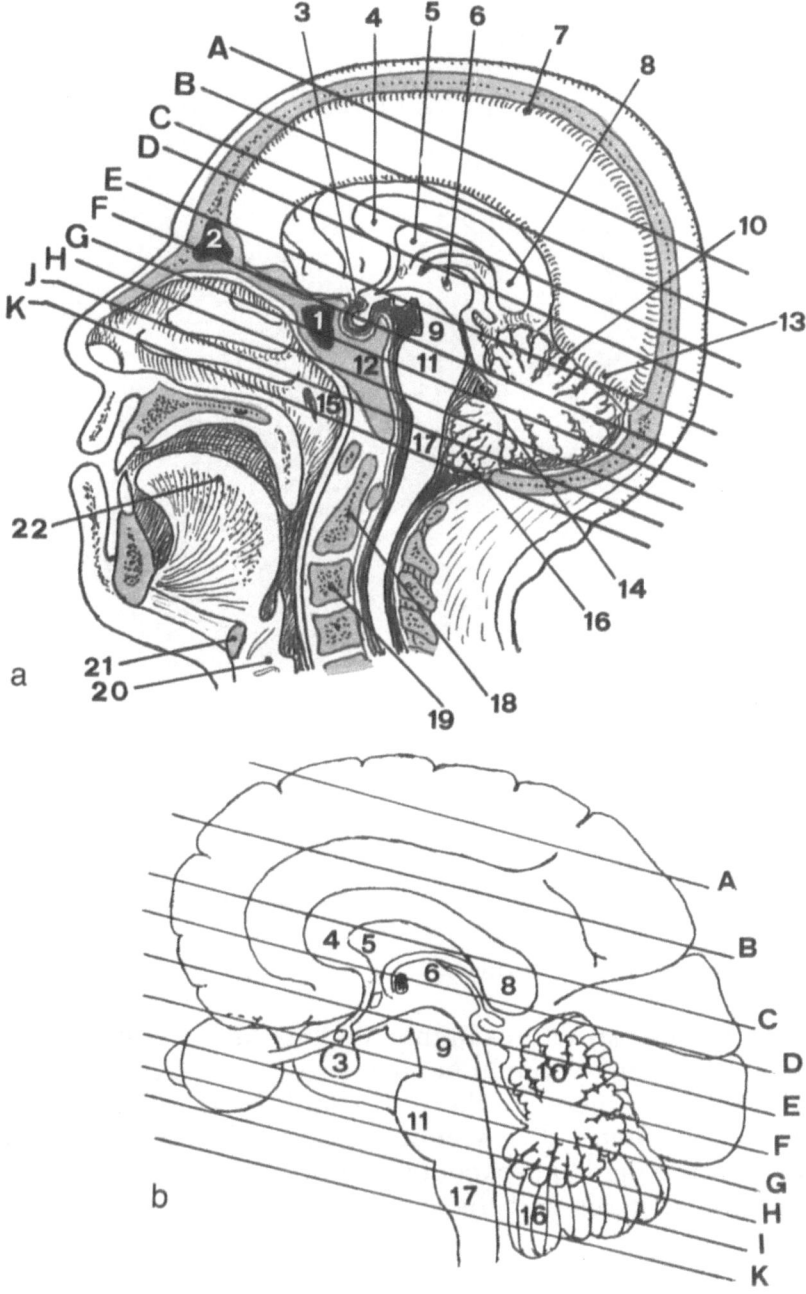

Fig. 11. Midsagittal section of head (*a*) and brain (*b*).

A–K Planes of section
1. Sphenoid sinus
2. Frontal sinus
3. Hypophysis (sella turcica)
4. Genu of corpus callosum
5. Septum pellucidum
6. Thalamus
7. Superior sagittal sinus

8. Splenium of corpus callosum
9. Mesencephalon
10. Cerebellar vermis
11. Pons
12. Clivus
13. Rectus sinus (tentorium cerebelli)
14. Fourth ventricle

15. Ostium of auditory tube
16. Cerebellar vermis
17. Medulla oblongata
18. Dens axis
19. Third cervical vertebra
20. Larynx
21. Hyoid bone
22. Tongue

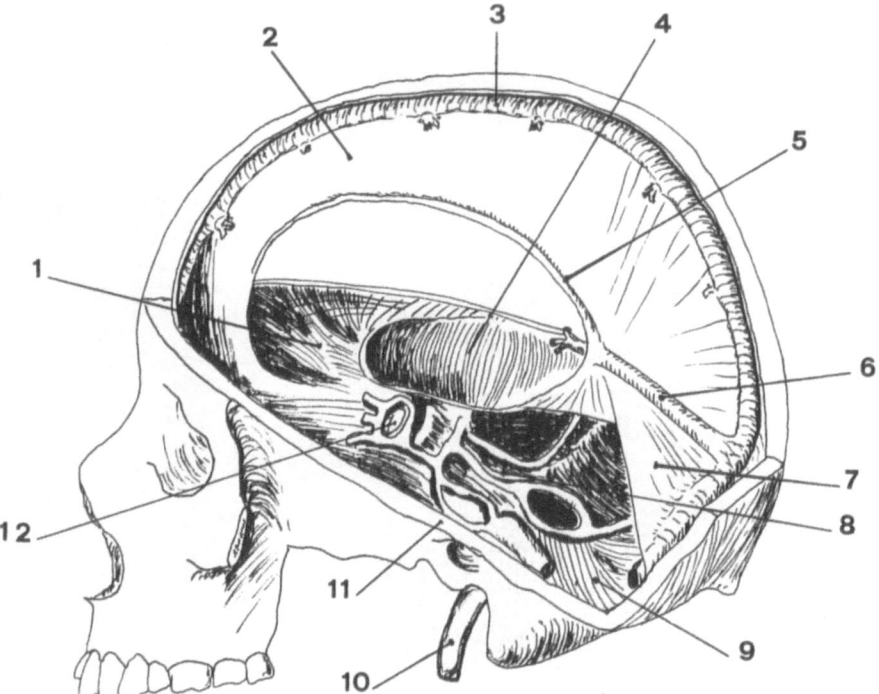

Fig. 12. Dura and dural sinuses.

1. Anterior cranial fossa	*6.* Rectus sinus	*11.* Temporal bone
2. Falx cerebri	*7.* Tentorium cerebelli	*12.* Sella turcica and cav-
3. Superior sagittal sinus	*8.* Clivus	ernous sinus
4. Middle cranial fossa	*9.* Posterior cranial fossa	
5. Inferior sagittal sinus	*10.* Internal jugular vein	

3.1.2 Horizontal Section A

The principal structures in this section (Fig. 13) are the cerebral hemispheres which are framed by the scalp and skull cap. Since this section passes through the medial aspect of the superior frontal gyrus and the paracentral lobule, the most conspicuous structures are gyri and sulci of the neocortex. The falx cerebri is seen in the midsagittal plane in the longitudinal cerebral fissure. Due to the arc-shaped trajectory of the superior sagittal sinus this structure is transected twice, anteriorly and posteriorly (Fig. 12).

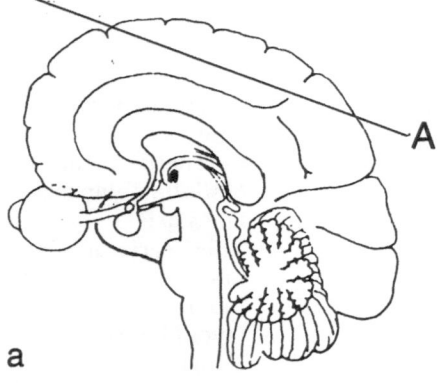

a

Fig. 13. *a–c.* Head: horizontal section A.

A Plane of section

1. Scalp		*8.* Dura mater
2. Skull cap		*9.* Internal lamina of
3. Superior sagittal		skull
sinus		*10.* Diploë of skull
4. Frontal lobe		*11.* External lamina of
5. Cerebral cortex		skull
6. Parietal lobe		*12.* Falx cerebri
7. Longitudinal cere-		
bral fissure		

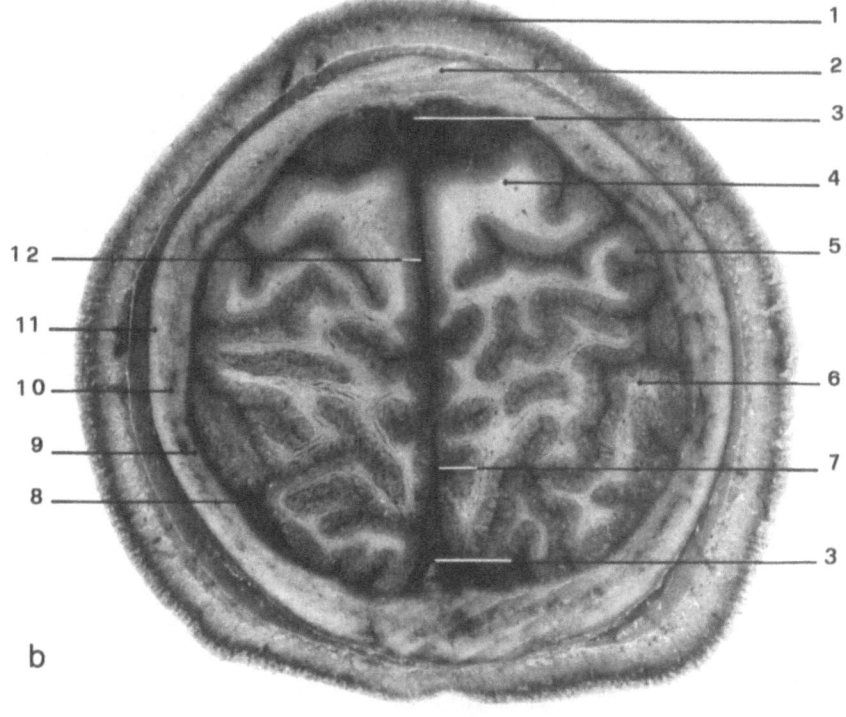

b

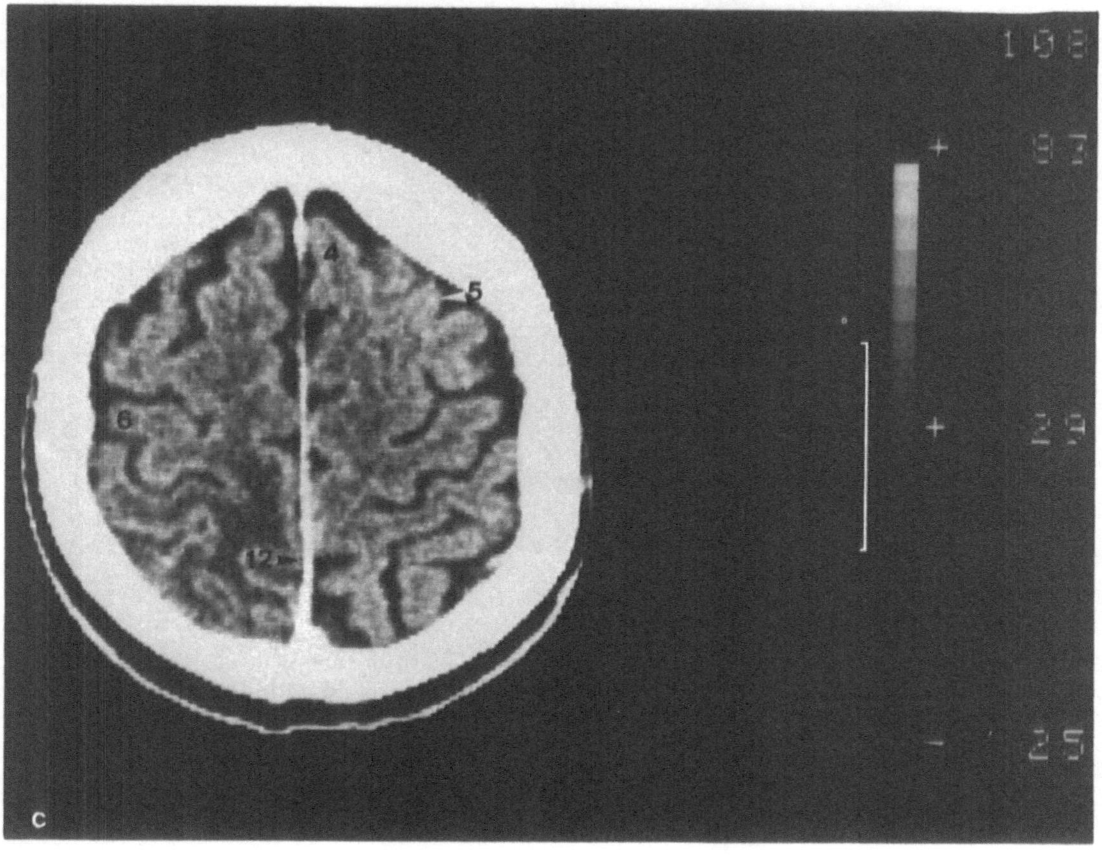

c

3.1.3 Horizontal Section B

This section (Fig. 14) lies halfway between the superior aspect of the cerebral cortex (Mantelkante) and the trunk of the corpus callosum. Principal structures at this level are the centrum semiovale and the cortical sulci and gyri. The falx cerebri is seen in the longitudinal cerebral fissure. The superior sagittal sinus is sectioned twice, anteriorly and posteriorly. Note that the cortex of the frontal pole is much thicker than that of the occipital pole.

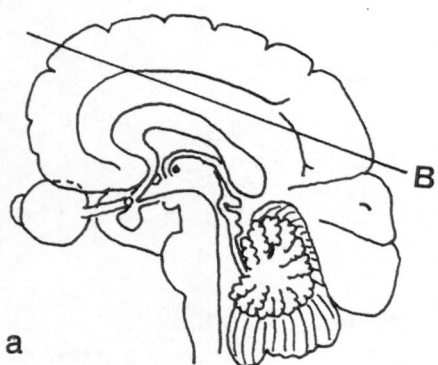

Fig. 14. *a–c.* Head: horizontal section *B.*

B Plane of section

1.	Scalp	*7.*	Centrum semiovale
2.	Superior sagittal sinus	*8.*	Galea aponeurotica
3.	Skull cap	*9.*	Occipital lobe
4.	Frontal lobe	*10.*	Parietal lobe
5.	Cerebral cortex	*11.*	Arachnoid
6.	Falx cerebri	*12.*	Diploë of skull

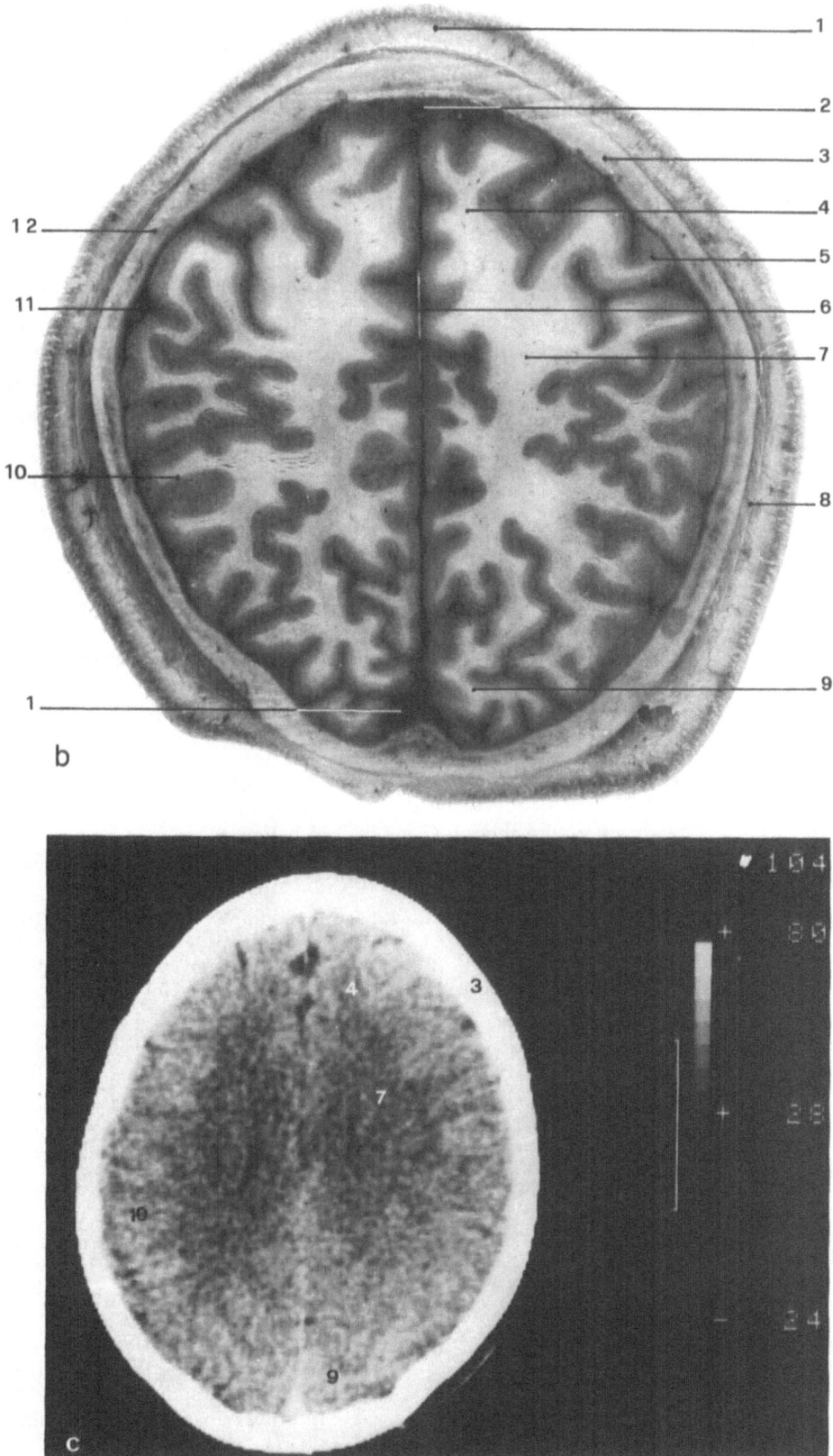

3.1.4 Horizontal Section C

The plane of section lies below the trunk of the corpus callosum (Figs. 15 and 16). Anteriorly it passes through the genu of the corpus callosum, posteriorly through the splenium of the corpus callosum. Lateral to the midline septum pellucidum one sees the frontal horns, the body, and the occipital horns of the lateral ventricles. Other visible midline structures are the anterior and posterior portions of the falx cerebri and the cut undersurface of the body of the corpus callosum. The caudate nucleus bulges into the lateral ventricle forming its lateral wall. The centrum semiovale is obvious on the left side. The right side is cut at a somewhat lower level and shows the beginning of the corpus striatum; the internal, external, and extreme capsules; the claustrum, and the insular cortex. The large paired frontal sinuses are clearly visible. In the CT image the calcified choroid plexuses of the lateral ventricles are well demonstrated (white structures).

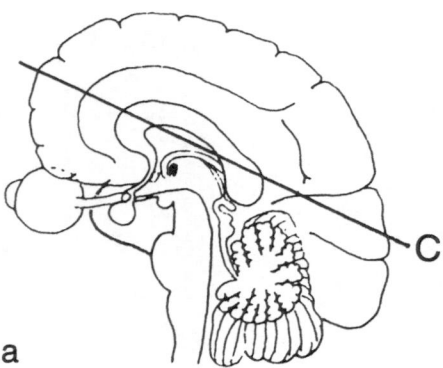

a

Fig. 15. *a–c.* Head: horizontal section *C*.

C Plane of section
1. Temporal lobe
2. Middle cerebral artery
3. Septum pellucidum
4. Lateral ventricle (frontal horn)

5. Corpus callosum (genu)
6. Anterior cerebral artery
7. Frontal sinus
8. Frontalis muscle
9. Frontal lobe
10. Circular sulcus of insula
11. Insular gyri
12. Caudate nucleus (body)
13. Claustrum
14. External capsule
15. Lateral ventricle (occipital horn)
16. Occipital lobe
17. Calcarine sulcus
18. Falx cerebri
19. Superior sagittal sinus
20. Inferior sagittal sinus
21. Posterior forceps of corpus callosum
22. Posterior cerebral artery
23. Corpus callosum (body)

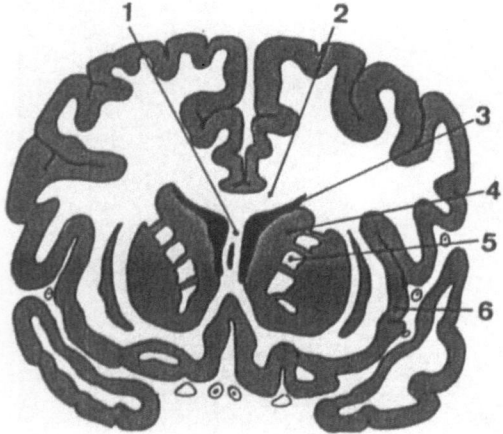

Fig. 16. Frontal section through brain at the level of the anterior limb of the internal capsule.

1. Septum pellucidum
2. Corpus callosum (body)
3. Lateral ventricle (frontal horn)
4. Caudate nucleus (head)
5. Internal capsule (anterior limb)
6. Insula

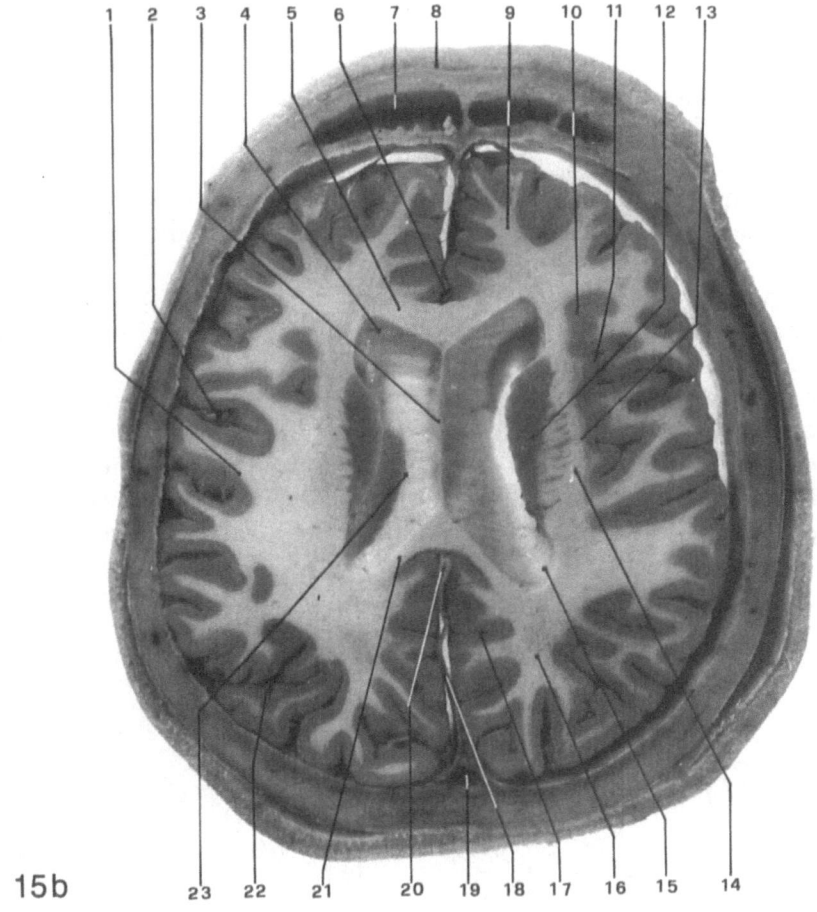

15b

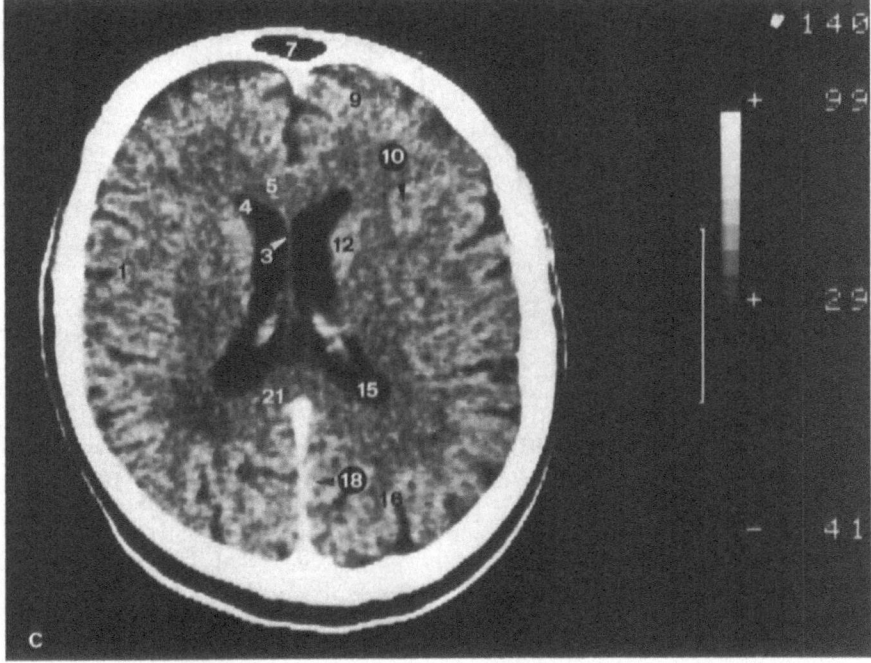

3.1.5 Horizontal Section *D*

The plane of section is through the paired thalami of the diencephalon, the great cerebral vein of Galen, and the superior aspect of the anterior cerebellar vermis (Fig. 17). In the median and paramedian zones the following important structures can be seen: the genu of the corpus callosum, the midline septum pellucidum which separates the two frontal horns of the lateral ventricles, the heads of the caudate nuclei which are the lateral borders of the frontal horns of the lateral ventricles, the two thalami, the crus and body of the right fornix, the trigone areas of the lateral ventricles containing the choroid plexuses, and the splenium of the corpus callosum. Lateral to the thalamus lie the internal capsule, the putamen, the external capsule, the claustrum, and the insular cortex. The optic radiations are seen as white bands found lateral to the trigones of the lateral ventricles. In the CT image the interventricular foramina of the lateral ventricles can be seen lateral to the columns of the fornix. The calcified pineal body is seen posterior to the third ventricle in the midline.

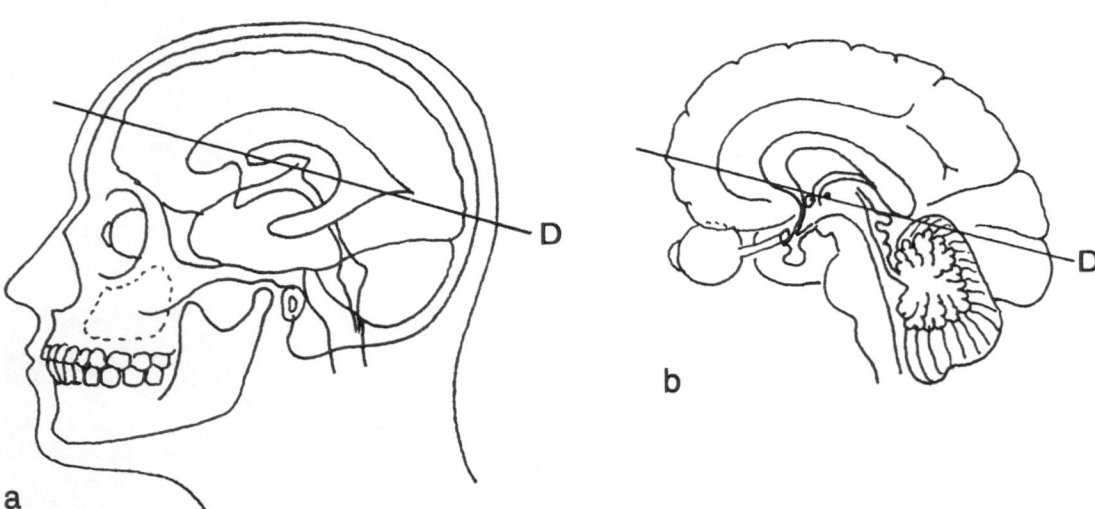

Fig. 17. *a–d.* Head horizontal section D.

D Plane of section

1. Lateral fissure
2. Insula
3. Thalamus
4. Caudate nucleus (head)
5. Corpus callosum (genu)
6. Frontal lobe
7. Septum pellucidum
8. Lateral ventricle (frontal horn)
9. Internal capsule (genu)
10. Putamen
11. Claustrum
12. Fornix
13. Choroid plexus of lateral ventricle
14. Corpus callosum (splenium)
15. Occipital lobe
16. Transverse sinus
17. Cerebellar vermis
18. Tentorium cerebelli
19. Great cerebral vein
20. Calcarine sulcus
21. Optic radiation
22. Temporal lobe
23. Insular gyrus

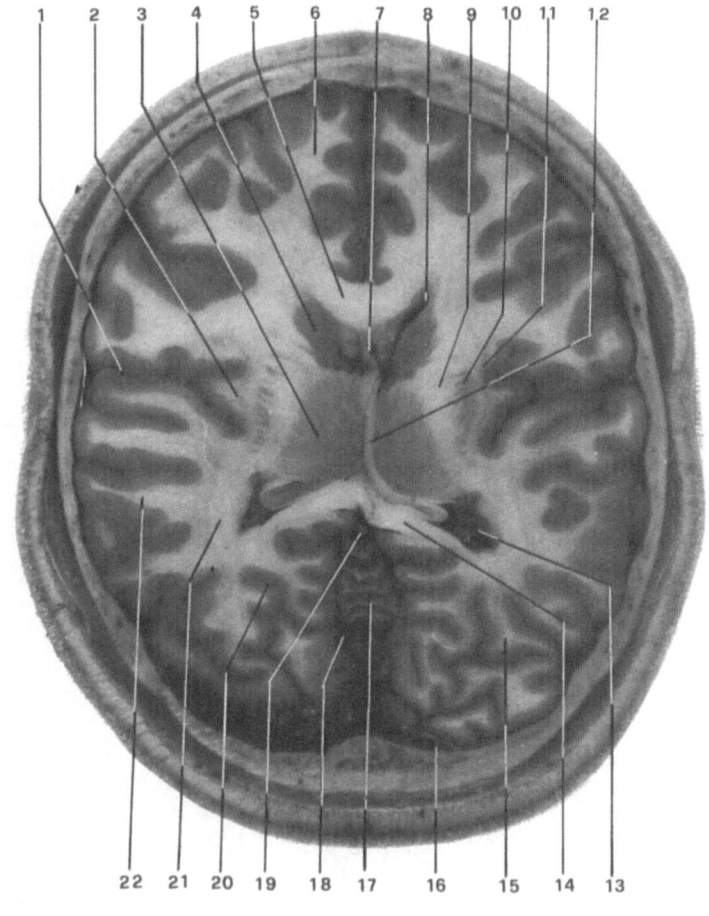

C

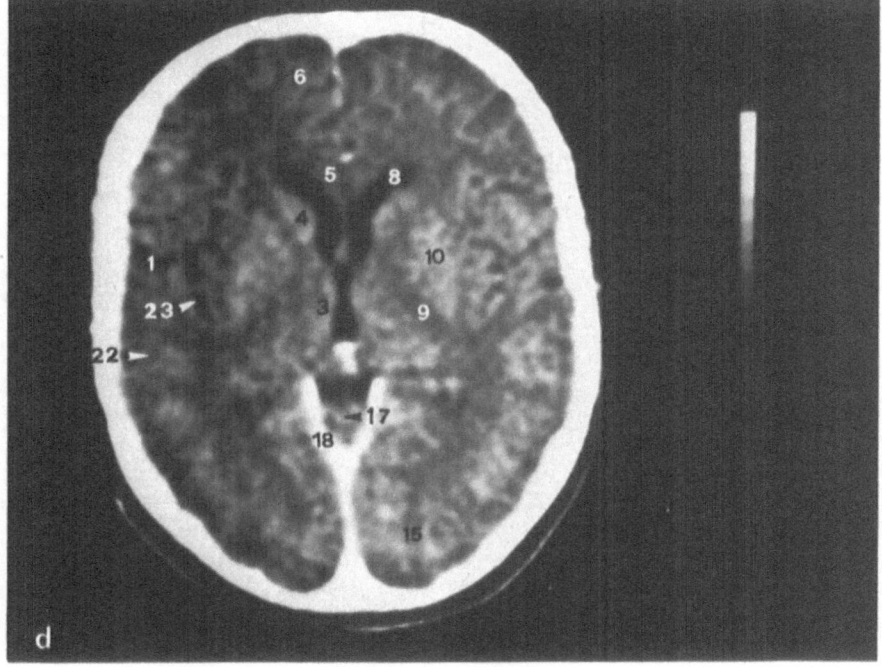

d

3.1.6 Horizontal Section *E*

The plane of section is through the midbrain (Fig. 18). Visible hemispheric structures are the frontal, temporal, and occipital lobes and the insula. In the midline the third ventricle, the midbrain, and the cerebeller vermis surrounded by the V-shaped tentorium cerebelli can be seen. At the apex of the tentorium lie the straight sinus and, occipital to this structure, the two initial segments of the transverse sinuses.

On the right side within the temporal lobe the temporal horn of the lateral ventricle and the hippocampus are prominent.

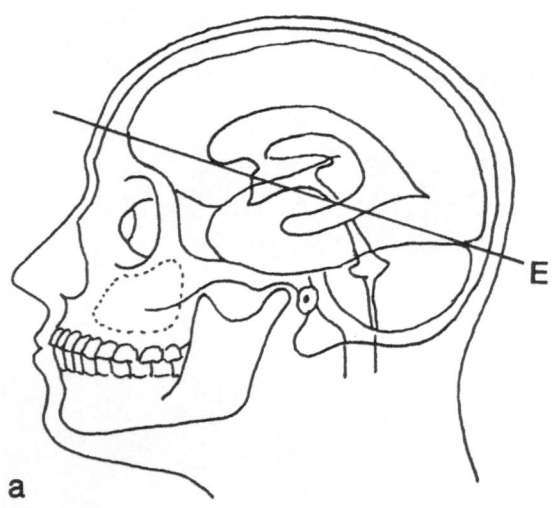
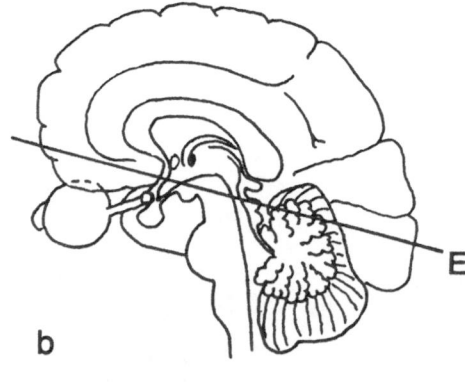
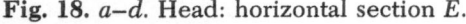

Fig. 18. *a–d.* Head: horizontal section *E.*

E Plane of section

1.	Temporal lobe	*16.*	Substantia nigra
2.	Lateral sulcus	*17.*	Mesencephalon
3.	Insula	*18.*	Cerebral aqueduct
4.	Claustrum	*19.*	Cerebellum
5.	Middle cerebral artery	*20.*	Rectus sinus
6.	Frontal lobe	*21.*	Internal occipital protuberance
7.	Anterior cerebral artery	*22.*	Transverse sinus
8.	Frontal sinus	*23.*	Occipital lobe
9.	Frontal crest	*24.*	Tentorium cerebelli
10.	Third ventricle	*25.*	Transverse cerebral fissure
11.	Nucleus accumbens septi	*26.*	Red nucleus
12.	Subarachnoid space	*27.*	Hippocampus
13.	Lesser wing of sphenoid bone	*28.*	Lateral ventricle (temporal horn)
14.	Temporalis muscle		
15.	Basis pedunculi		

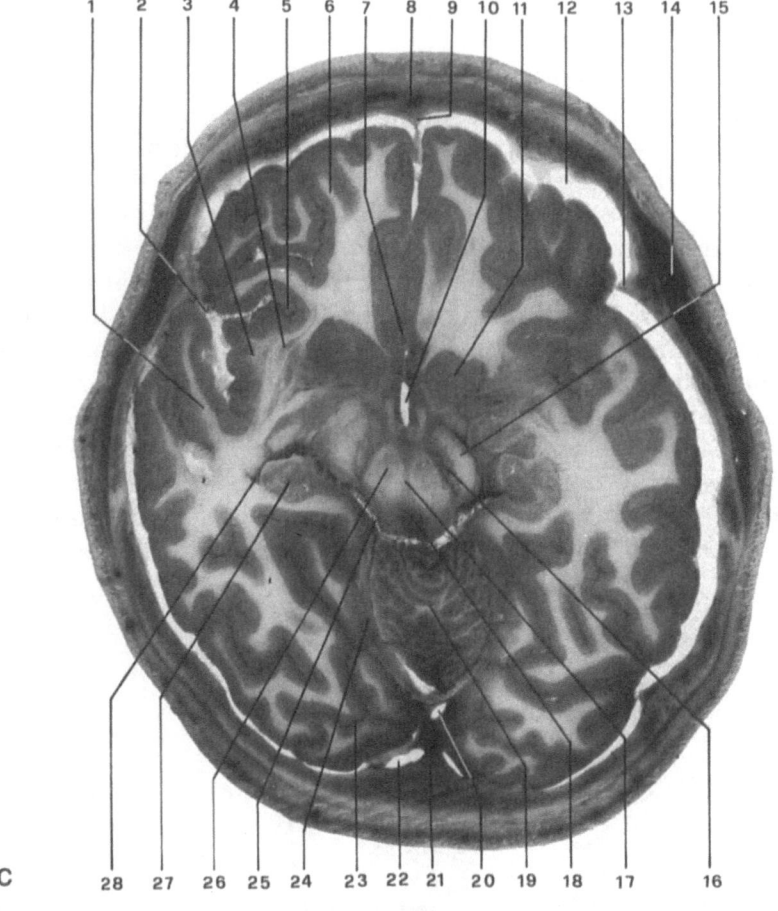

C

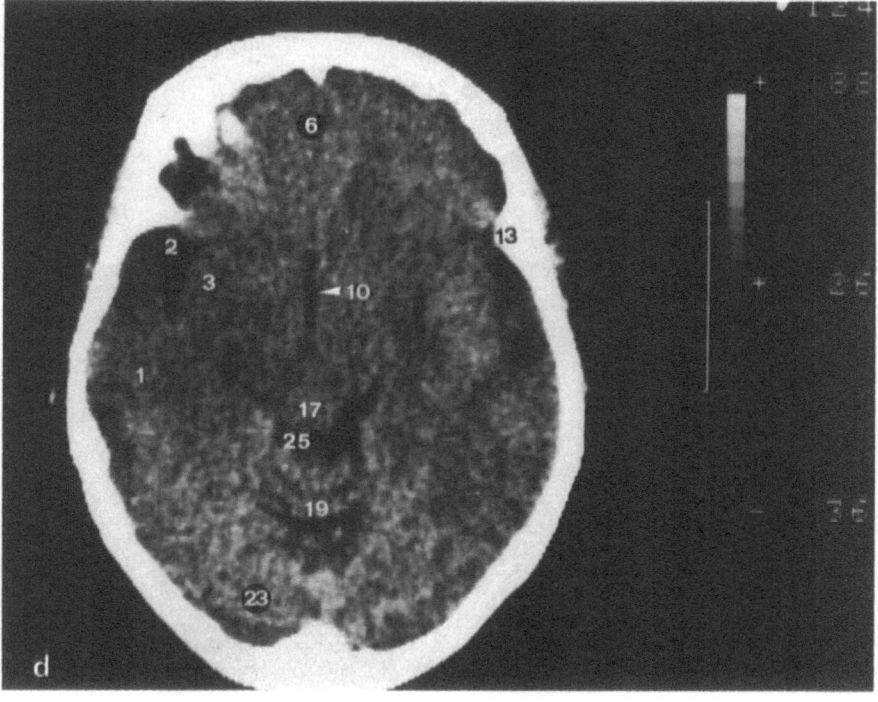

d

3.1.7 Horizontal Section *F*

The plane of section is through the superior orbit, the three cranial fossae, the optic chiasm, and the midbrain at the level of the inferior colliculi (Fig. 19). Large paired frontal sinuses are present in the frontal bone. In the orbit the levator palpebrae superioris muscle and orbital fat are seen. Structures of the anterior cerebral fossa include the frontal lobe and its gyrus rectus, olfactory sulcus, and orbital gyri. The anterior cerebral arteries are in the longitudinal cerebral fissure. The middle cerebral fossae contain the temporal lobes. The two unci of the temporal lobes are seen in their important relationships to the cerebral peduncles and the oculomotor nerves of the midbrain. Also in the central part of the middle cranial fossa are the optic chiasm, the diaphragma sellae, and the internal carotid arteries. The middle cerebral arteries are visible in the lateral cerebral fissures. The midbrain lies in the tentorial notch of the posterior fossa. The basilar artery lies posterior to the dorsum sellae. In the posterior fossa the cerebellum is bounded laterally by the tentorium cerebelli and posteriorly by the transverse sinuses, the confluence of the sinuses, and the internal occipital protuberance.

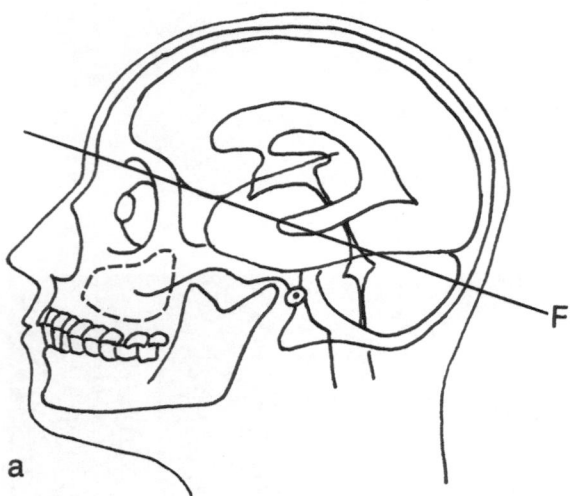

Fig. 19. *a–d.* Head: horizontal section *F.*

1. Diaphragma sellae	*13.* Optic chiasm
2. Internal carotid artery	*14.* Temporalis muscle
3. Middle cerebral artery	*15.* Temporal lobe
4. Temporal pole	*16.* Oculomotor nerve
5. Levator palpebrae superioris muscle	*17.* Basis pedunculi
6. Nasofrontal vein	*18.* Mesencephalon
7. Frontal sinus	*19.* Cerebral aqueduct
8. Olfactory sulcus	*20.* Confluence of sinuses
9. Crista galli	*21.* Cerebellum
10. Frontal lobe	*22.* Transverse sinus
11. Anterior cerebral artery	*23.* Tentorium cerebelli
12. Orbital fat	*24.* Sigmoid sinus
	25. Basilar artery

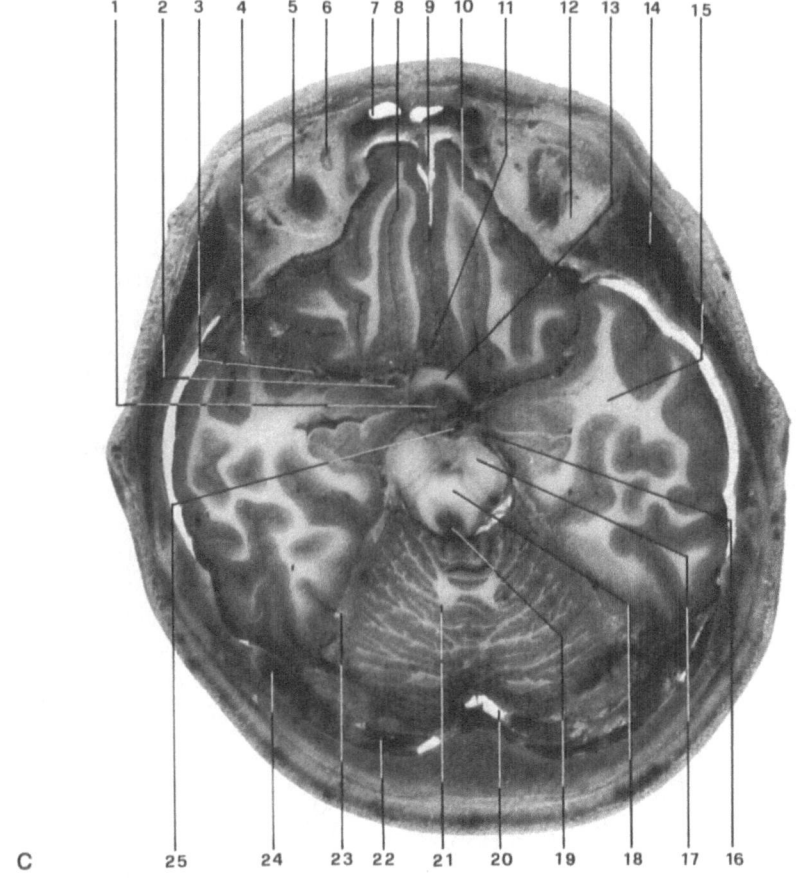

C

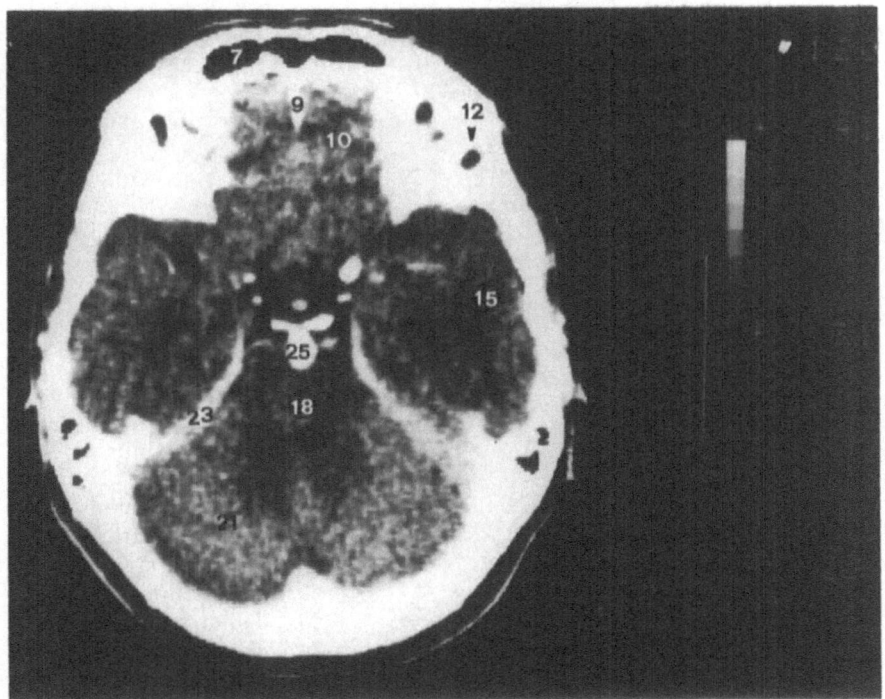

d

3.1.8 Horizontal Section G

The plane of section passes through the orbit and eyeball, the frontal sinuses, the roof of the nasal cavity, the olfactory tract, the hypophysis, and the internal occipital protuberance (Fig. 20). The following orbital and related structures can be seen: lacrimal glands, retrobulbar fat, the eyeballs, levator palpebrae superioris muscle, superior oblique muscle, and the optic nerve at the orbital apex.

Through the sectioned roof of the nasal cavity one sees the gyrus rectus, olfactory sulcus and medial orbital gyrus. The temporal lobes are in the middle cranial fossae. Important visible structures of the sellar region are the hypophysis, internal carotid artery, cavernous sinus, and anterior clinoid process. In the posterior fossa we see the basilar artery, pons, and cerebellum with its dentate nuclei located in the corpus medullare. The two superior cerebellar peduncles flank the rostral aditus of the fourth ventricle. In the CT image the petrous portions of the temporal bones are visible.

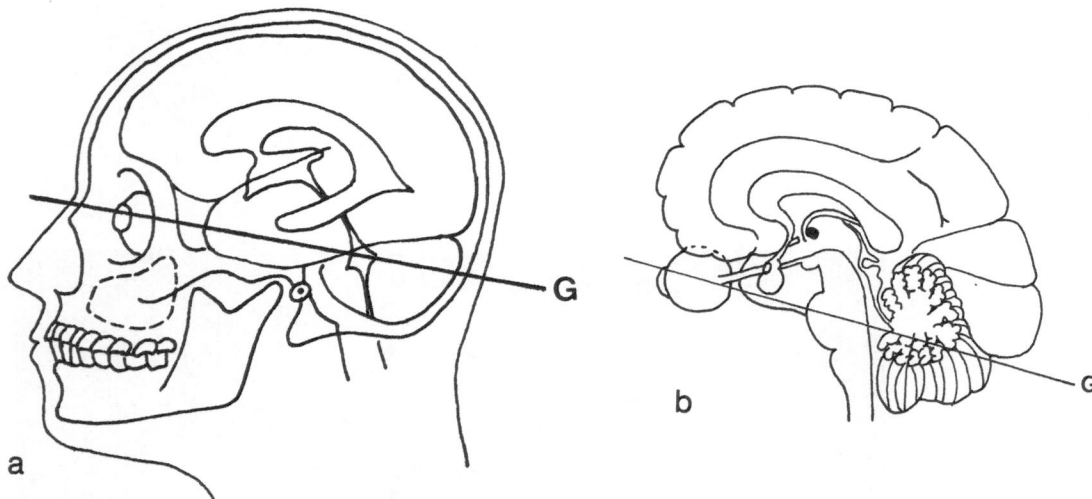

Fig. 20. *a–d.* Head: horizontal section *G*.

1.	Cavernous sinus	*13.*	Optic nerve
2.	Cyst in temporal pole	*14.*	Hypophysis
3.	Internal carotid artery	*15.*	Temporalis muscle
4.	Lacrimal gland	*16.*	Basilar artery
5.	Orbital fat	*17.*	Sigmoid sinus
6.	Levator palpebrae superioris muscle	*18.*	Pons
7.	Superior oblique muscle	*19.*	Dentate nucleus
8.	Crista galli	*20.*	Fourth ventricle
9.	Frontal sinus	*21.*	Occipital sinus
10.	Olfactory tract	*22.*	Cerebellum
11.	Frontal lobe (inferior surface)	*23.*	Tentorium cerebelli
12.	Bulbus oculi	*24.*	Temporal lobe
		25.	Temporal bone, petrous portion

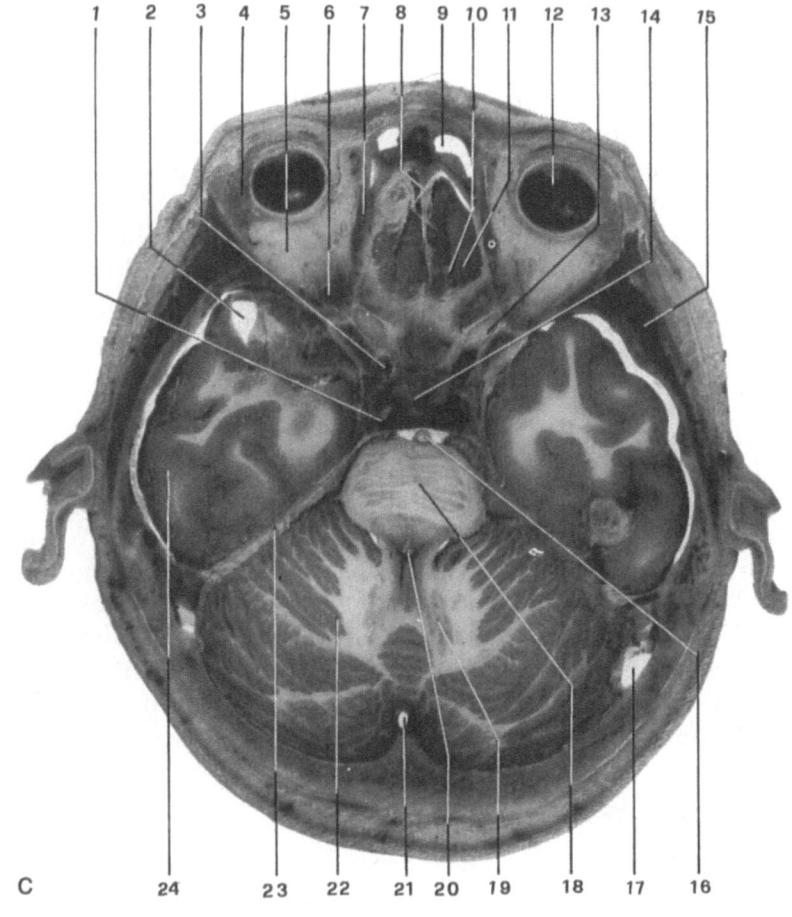

C

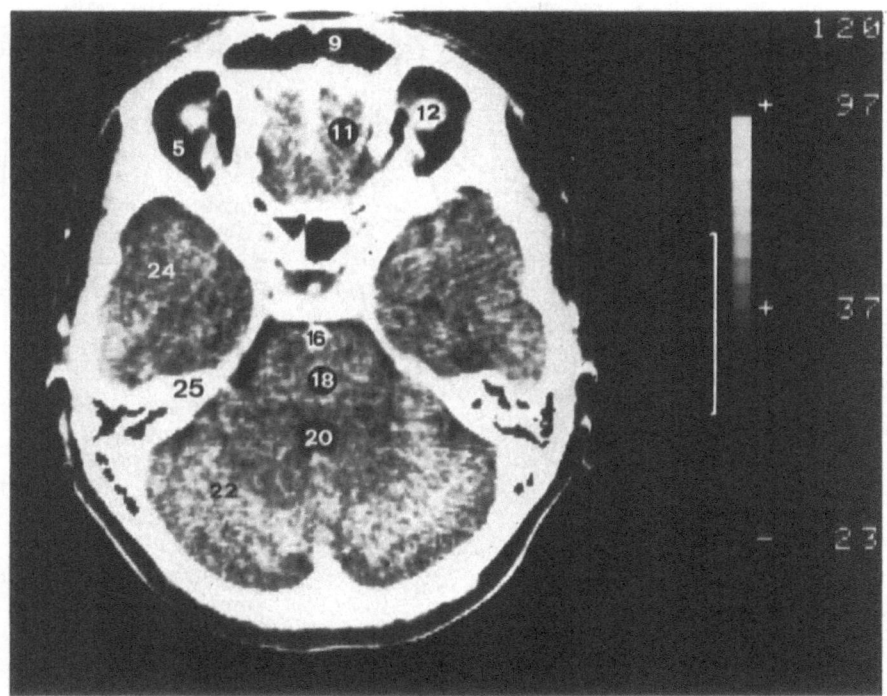

d

3.1.9 Horizontal Section H

The plane of section is through the middle portion of the orbit, above the floor of the middle cranial fossa, cutting through the petrous portion of the temporal bone and passing through the posterior cranial fossa (Fig. 21). The visceral cranium is represented by the two orbits and the midline nasal cavity. Within the orbit one sees the eyeball and the attached lateral and medial recti muscles. Within the muscular cone lie the optic nerve and retrobulbar fat. Anterior to the globe are the upper eyelid and the conjunctival fornix. The ethmoid air cells represent the lateral walls of the nasal cavity; the posterior wall consists of the sphenoid sinuses (note the asymmetry).

In the middle cranial fossa the inferior aspect of the temporal lobe is visible laterally, the body of the sphenoid bone lies in the midline, and the carotid siphon is seen in the cavernous sinus. The petrous portion of the temporal bone is the boundary between the middle and posterior cranial fossae. The anteromedial portion of the petrosal bone houses the membranous labyrinth; the posterolateral portion contains the mastoid air cells. In the posterior cranial fossa one sees the midline pons connected to the cerebellar hemispheres by the two middle cerebellar peduncles. The cerebellar vermis lies in the posterior cerebellar incisure. In the midline between the pontine tegmentum and the cerebellar vermis lies the fourth ventricle. The largest of the deep cerebellar nuclei, the dentate nucleus, lies in the centrum ovale cerebelli.

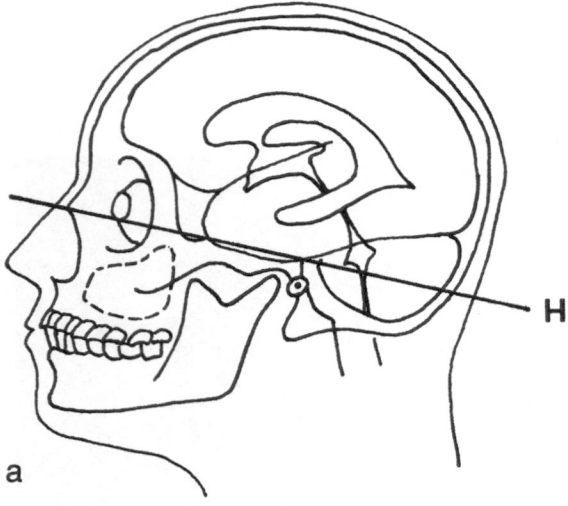

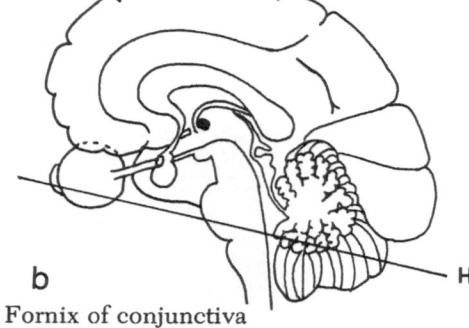

b

H

H

a

Fig. 21. *a–d*. Head: horizontal section *H*.

1. Temporal lobe
2. Temporalis muscle
3. Lateral rectus muscle
4. Upper eyelid
5. Optic nerve
6. Medial rectus muscle
7. Nasal septum
8. Sphenoid sinus
9. Ethmoid air cells
10. Body of sphenoid bone
11. Internal carotid artery (siphon)

12. Fornix of conjunctiva
13. Trigeminal ganglion
14. Trigeminal nerve
15. Petrous portion of temporal bone
16. Tympanic cavity
17. Mastoid air cells
18. Sigmoid sinus
19. Middle cerebellar peduncle
20. Dentate nucleus
21. Cerebellar tonsil
22. Internal occipital protuberance
23. Occipital sinus
24. Trapezius muscle
25. Semispinalis capitis muscle
26. Fourth ventricle
27. Pons
28. Cerebellum
29. Posterior semicircular canal
30. Basilar artery

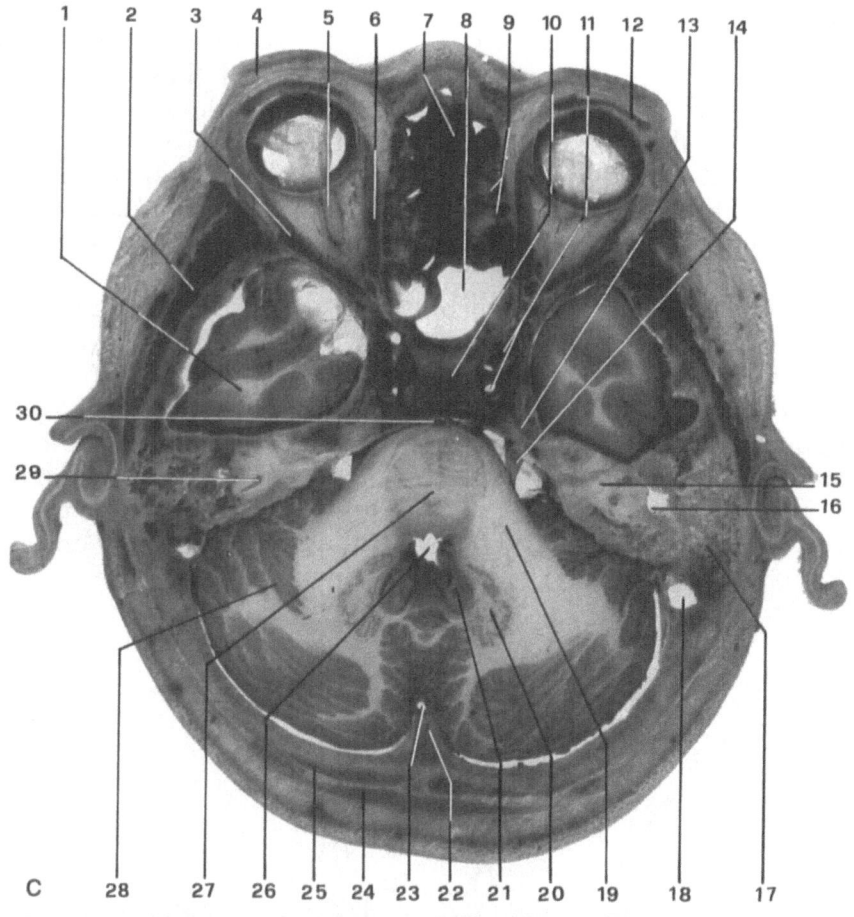

C

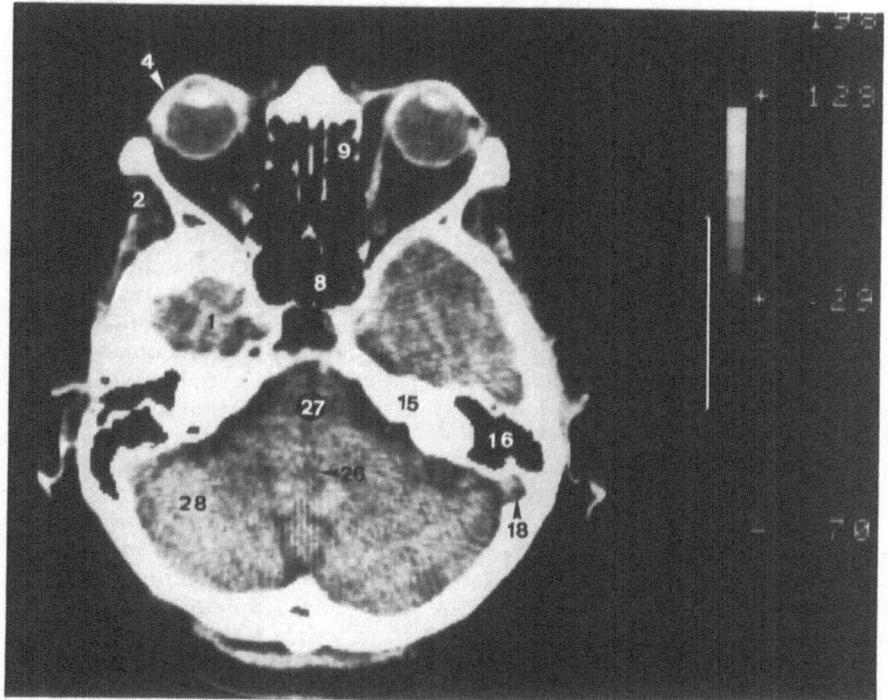

d

3.1.10 Horizontal Section *I*

The plane of section lies slightly above the floor of the orbit, passing through the infratemporal fossa, the basal part of the petrous portion of the temporal bone and occipital bone, and the posterior cranial fossa above its floor (Fig. 22). Within the orbit the cut surfaces of the inferior rectus and the inferior oblique muscles are visible in the orbital fat. The maxillary sinuses appear between the orbital and nasal cavities.

The muscles of mastication (temporalis and lateral pterygoid) and the temporomandibular joint are in the infratemporal fossa. In the midline the section passes through the base of the skull (basilar portion of the occipital bone). One sees the cartilaginous part of the external acoustic meatus, the mastoid air cells in the petrous part of the temporal bone, and at the apex of the pyramid, the carotid canal and the cartilaginous portion of the auditory tube. The jugular foramen and its conduit, the sigmoid sinus, are located between the occipital bone and the petrous portion of the temporal bone. The medulla oblongata and the vertebral arteries lie on the clivus. The inferior surface of the cerebellum lies in the posterior cranial fossa just above its floor. The neck muscles are seen external to the occipital bone.

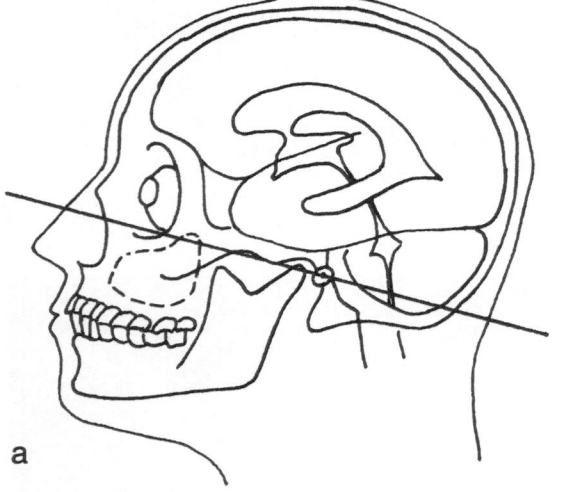

a

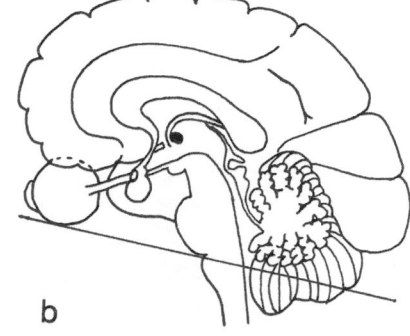

b

Fig. 22. *a–d.* Head: horizontal section *I*.

1. Vomer
2. Superior nasal concha
3. Inferior oblique muscle
4. Inferior fornix of conjunctiva
5. Inferior rectus muscle
6. Orbital fat
7. Lacrimonasal duct
8. Perpendicular lamina of ethmoid bone
9. Nasal septum
10. Maxillary sinus
11. Sphenoid sinus
12. Lateral pterygoid muscle
13. Temporalis muscle
14. Masseter muscle
15. Zygomatic arch
16. Articular disc
17. Head of mandible
18. Internal carotid artery
19. Mastoid air cells
20. Sigmoid sinus
21. Jugular foramen
22. Medulla oblongata
23. Central canal (closed medulla)
24. Nuchal ligament
25. Cerebellomedullary cistern
26. Rectus capitis (posterior minor) muscle
27. Semsispinalis capitis muscle
28. Trapezius muscle
29. Cerebellum
30. Splenius capitis muscle
31. Vertebral artery
32. External auditory meatus
33. Occipital bone (pars basilaris)
34. Auditory tube (pars cartilaginea)

1 2 3 4 5 6 7 8 9 10 11 12 13 14 15
34
33
32
31
16
17
18
30 29 28 27 26 25 24 23 22 21 20 19

c

7
9
6
10
2
1
15
17
33
32
22
19
25
29

d

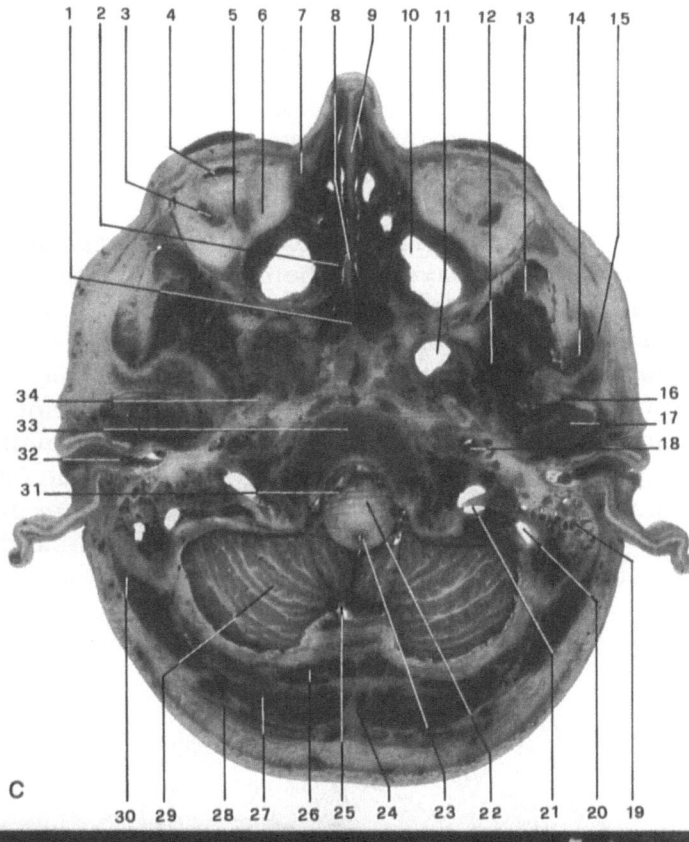

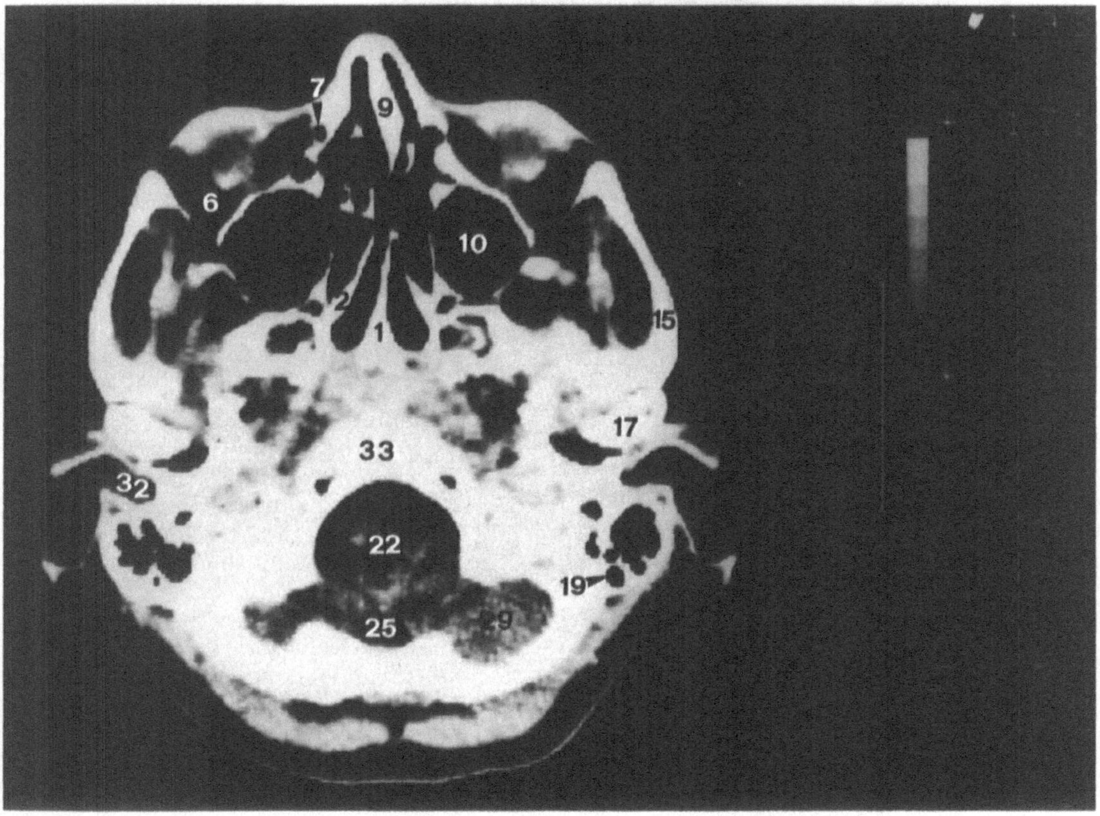

3.1.11 Horizontal Section *K*

The plane of section (Fig. 23) is through the maxillary sinus and the infratemporal fossa and is tangent to the inferior surface of the occipital bone at the level of the foramen magnum (Fig. 9). In the facial area the musculature is seen as a gray layer just beneath the skin. In the nasal cavity the section transects the septum, the middle concha, and on the lateral wall the lacrimonasal duct. Posterior to the maxillary sinus and within the infratemporal fossa one can see the masseter, temporalis, and lateral pterygoid muscles, the coronoid process and the collum of the mandible, and laterally the parotid gland. On the inferior surface of the base of the skull is the epipharynx; lateral; the auditory tube; and posterior to the auditory tube, the longus capitis anterior muscle.

Within the foramen magnum lies the medulla oblongata, the vertebral arteries, and the cerebellar tonsils. Lateral to the foramen magnum are the occipital condyles, posterior to the condyles the vertebral arteries. The insertions of the neck muscles are seen on the inferior surface of the squama of the occipital bone. Lateral to the squama of the occipital bone, the mastoid processes are visible. The internal jugular vein and the internal carotid artery lie lateral to the occipital condyle.

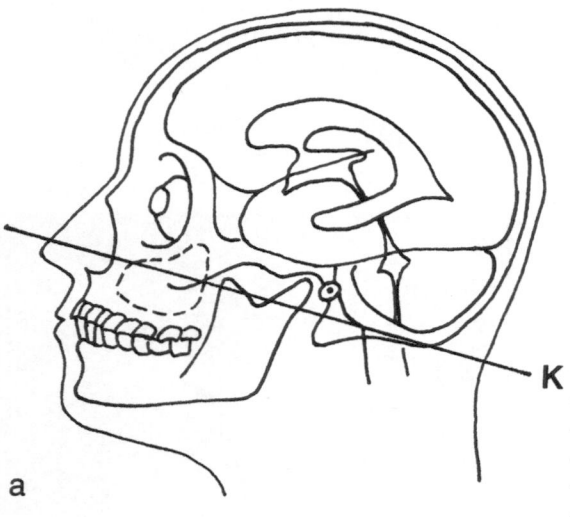

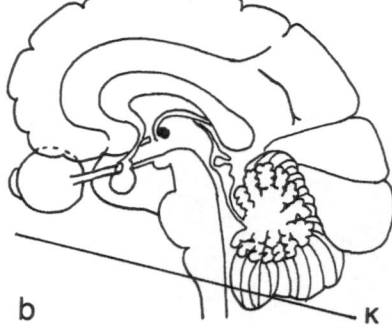

Fig. 23. *a–d.* Head: horizontal section *K*.

1. Masseter muscle
2. Tensor veli palatini muscle
3. Temporalis muscle
4. Auditory tube
5. Maxillary sinus
6. Choana
7. Lacrimonasal duct
8. Nasal septum
9. Inferior nasal concha
10. Nasal cavity
11. Perpendicular lamina of ethmoid bone
12. Epipharynx
13. Zygomatic tubercle
14. Coronoid process of mandible
15. Neck of mandible
16. Parotid gland
17. Internal carotid artery
18. Internal jugular vein
19. Mastoid process
20. Occipital condyle
21. Vertebral artery (in sulcus of atlas)
22. Vertebral artery
23. Tonsil of cerebellum
24. Foramen magnum
25. Nuchal ligament
26. Rectus capitis anterior muscle
27. Semispinalis capitis muscle
28. Trapezius muscle
29. Splenius capitis muscle
30. Medulla oblongata
31. Sternocleidomastoid muscle
32. Mastoid air cells
33. Longus capitis anterior muscle
34. Levator veli palatini muscle
35. Lateral pterygoid muscle

1 2 3 4 5 6 7 8 9 10 11 12 13 14 15
16
17
18
35
34
33
32 31 30 29 28 27 26 25 24 23 22 21 20 19

c

200
130
30
-70
10
5
13
3
14
11
6
1
35
4
15
20
32
30
24
19

d

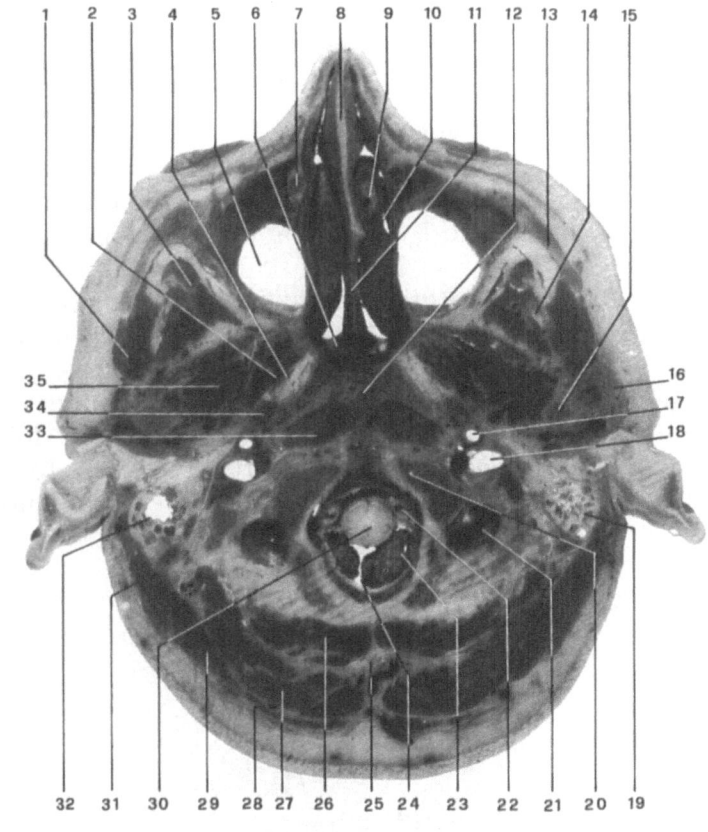

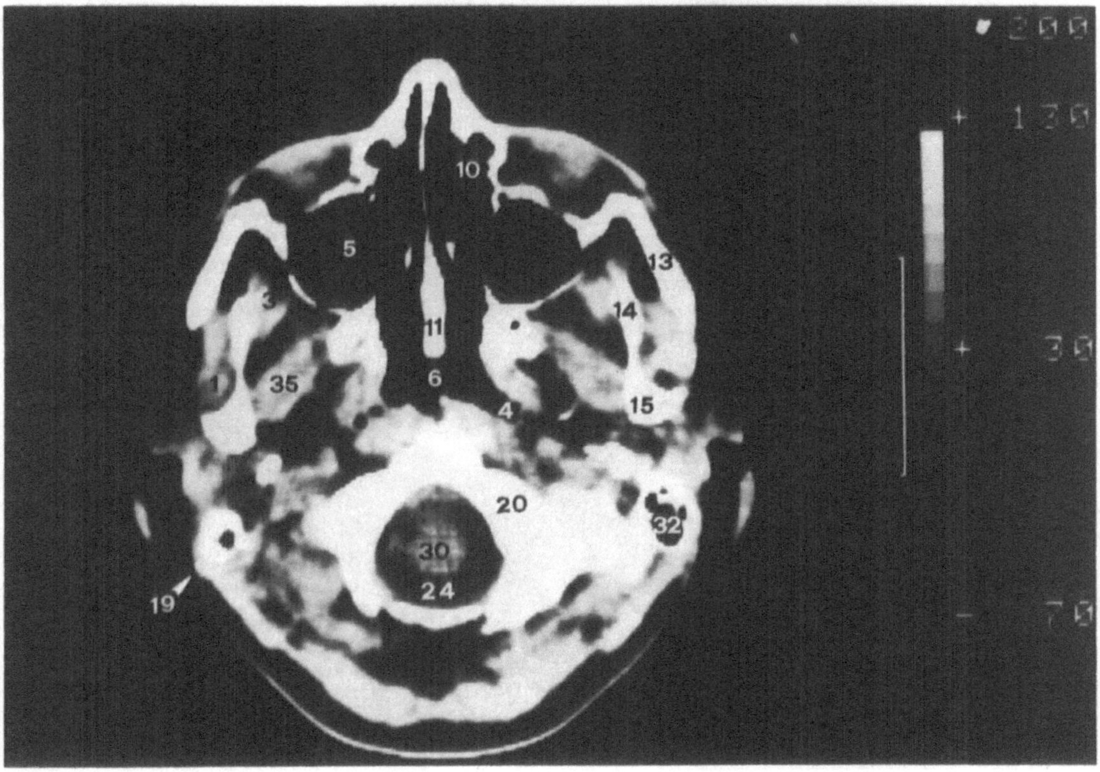

3.2 Neck

3.2.1 General Overview

The topography of the neck region can best be reviewed by studying a transverse section, a frontal schematic drawing, and a lateral schematic drawing of this area.

Figure 24 represents a transverse section through the middle third of the neck; the viscera lie ventral, the vertebral column and neck muscles lie dorsal. The cervical vertebrae contain foramina transversaria which conduct the vertebral arteries and veins. The emerging spinal nerves lie dorsal to these vessels. The vertebral column is surrounded by musculature. Prevertebral muscles are the longus colli and rectus capitis anterior. Laterovertebral muscles are the scaleni and levatores scapulae.

The deep (intrinsic) muscles of the neck lie lateral to the spinous processes of the vertebrae and are covered dorsally by the trapezius muscle. The longitudinally disposed prevertebral and deep muscles of the neck

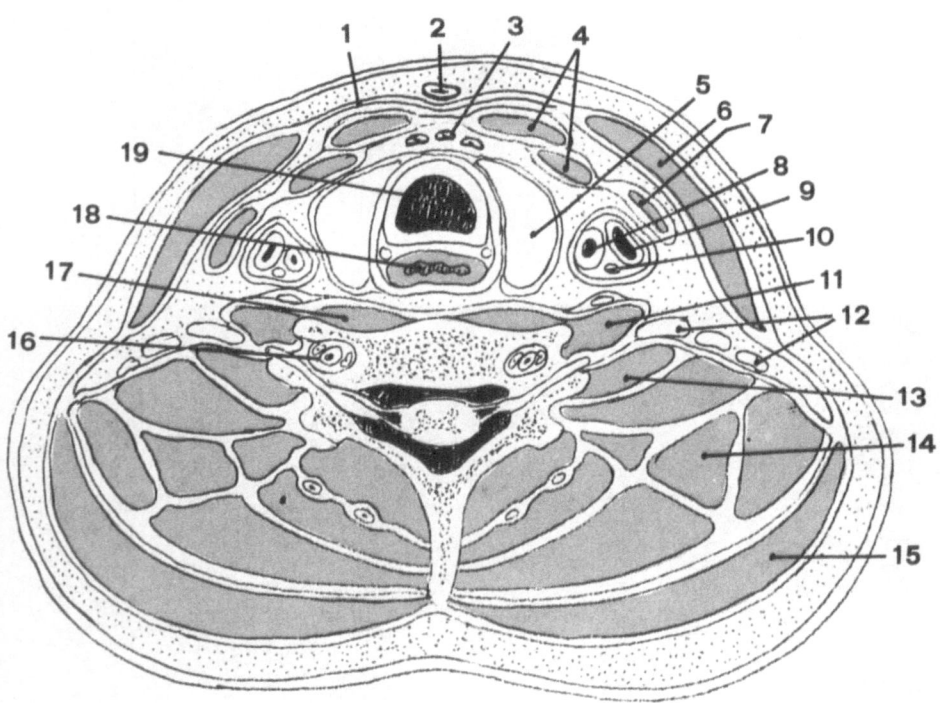

Fig. 24. Cross-section of neck (schematic).

1. Superficial cervical fascia
2. Anterior jugular vein
3. Inferior thyroid veins
4. Infrahyoid muscles
5. Thyroid gland
6. Sternocleidomastoid muscle
7. Omohyoid muscle
8. Common carotid artery
9. Internal jugular vein
10. Vagal nerve
11. Scalenus anterior muscle
12. Brachial plexus
13. Sclenus medius muscle
14. Neck musculature
15. Trapezius muscle
16. Vertebral artery and vein
17. Longus colli muscle
18. Esophagus
19. Trachea

maintain their position throughout the height of section in the transverse plane. The conically disposed scaleni move laterally as the level of section descends from cranial to caudal. The trapezius muscle becomes broader as the height of section moves caudal.

In the suprahyoid region the viscera are represented by the pharynx. Inferior to the hyoid bone the viscera of the neck can be subdivided into four parts: the larynx lies ventral and cranial, the trachea lies ventral and caudal, the laryngeal part of the pharynx lies dorsal and cranial, and the esophagus lies dorsal and caudal (Fig. 26). The larynx and cranial portion of the trachea are covered by the horseshoe-shaped thyroid gland (Fig. 24). The carotid sheath lies lateral to the visceral organs. Inferior to the hyoid bone this sheath contains the common carotid artery and the internal jugular vein;

between and dorsal to these vessels lies the vagal nerve. Superior to the hyoid bone the common carotid artery bifurcates into the internal and external carotid arteries (Figs. 25 and 26). The viscera and carotid sheath are surrounded by muscles. The sternocleidomastoid muscle is laterally and superficially disposed, the infrahyoid muscles are ventral, and the omohyoid muscle is lateral and deep to the sternocleidomastoid (Fig. 24).

The infrahyoid muscles maintain their position at all levels. Due to the oblique course of the sternocleidomastoid and omohyoid muscles, their positions change with respect to the level of section (Fig. 25 and 26).

The trunks of the brachial plexus and their relationships to the scalenus muscles are shown in Figs. 25 and 26. The gross anatomical relationships of the larynx are also shown in Figs. 25 and 26.

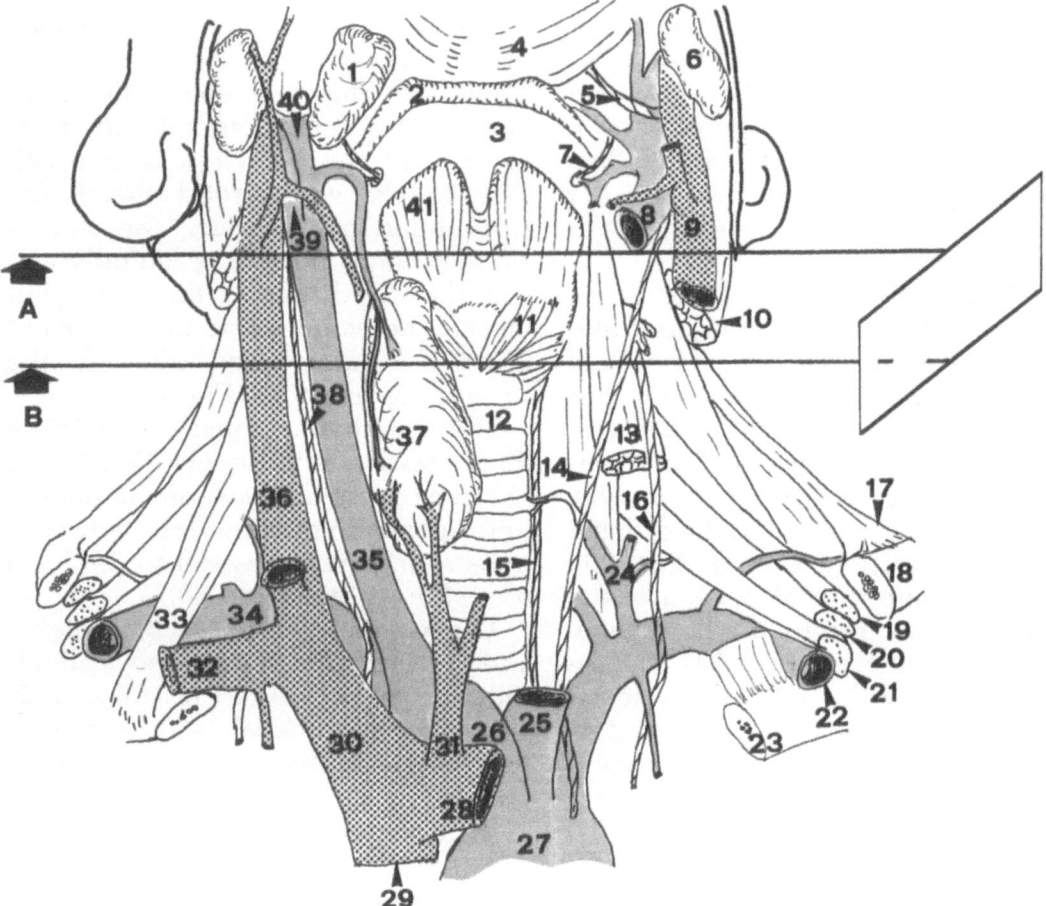

Fig. 25. Anterior aspect of neck.

A and B Planes of section

1. Submandibular gland
2. Hyoid bone
3. Thyrohyoid membrane
4. Raphe of mylohyoid muscle
5. Hypoglossal nerve
6. Parotid gland
7. Superior laryngeal nerve
8. Common carotid artery
9. Internal jugular vein
10. Sternocleidomastoid muscle
11. Cricothyroid muscle
12. Trachea
13. Scalenus anterior muscle
14. Left vagal nerve
15. Recurrent laryngeal nerve
16. Phrenic nerve
17. Trapezius muscle
18. Clavicle
19. Superior trunk, brachial plexus
20. Medial trunk, brachial plexus
21. Inferior trunk, brachial plexus
22. Subclavian artery
23. First rib
24. Thyrocervical trunk
25. Left common carotid artery
26. Brachiocephalic trunk
27. Aortic arch
28. Left brachiocephalic vein
29. Superior vena cava
30. Right brachiocephalic vein
31. Inferior thyroid vein
32. Right subclavian vein
33. Scalenus anterior muscle
34. Right subclavian artery
35. Right common carotid artery
36. Right internal jugular vein
37. Thyroid gland
38. Vagal nerve
39. Right internal carotid artery
40. Right external carotid artery
41. Thyroid cartilage

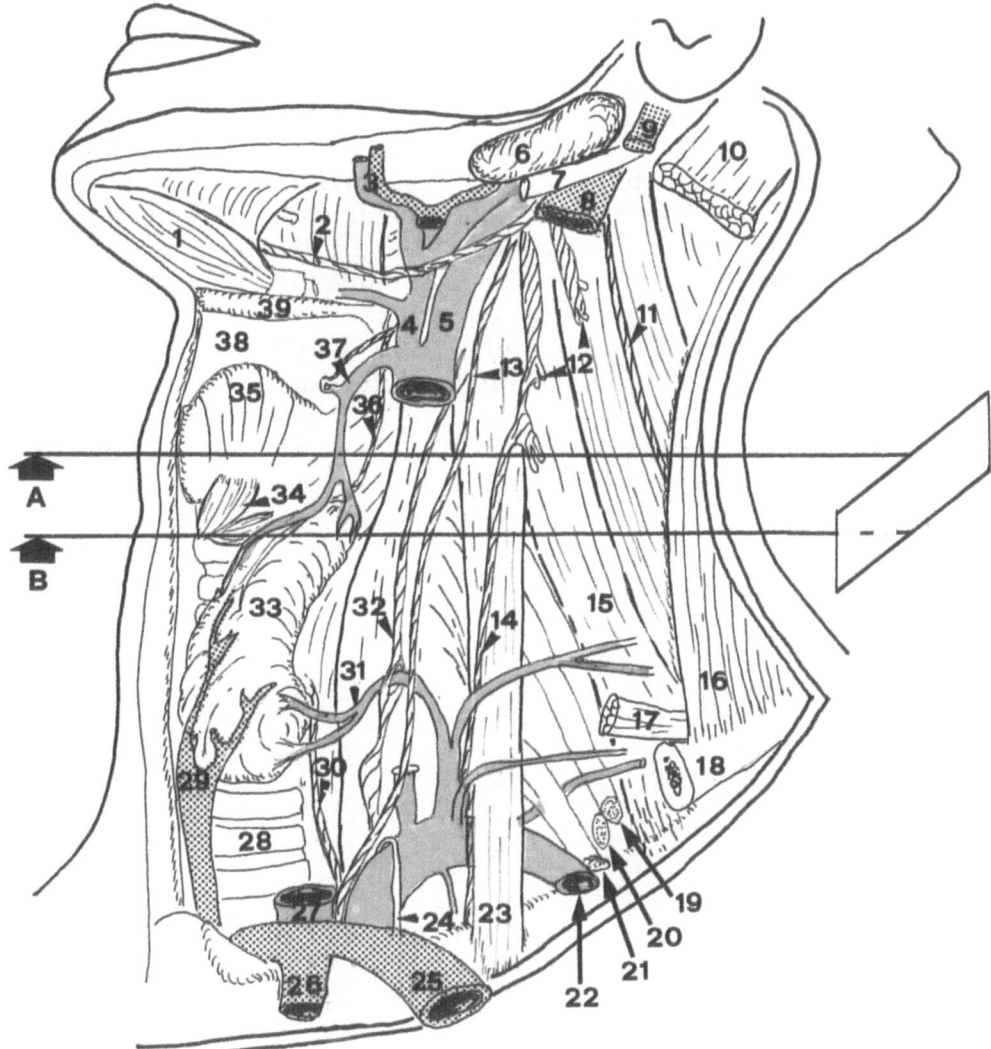

Fig. 26. Lateral aspect of neck.

A and *B* Planes of section

1. Anterior belly of digastric muscle
2. Hypoglossal nerve
3. Anterior facial vessels
4. External carotid artery
5. Internal carotid artery
6. Submandibular gland
7. Posterior belly of digastric muscle
8. Internal jugular vein
9. External jugular vein
10. Sternocleidomastoid muscle
11. Accessory nerve
12. Cervical nerves
13. Vagal nerve
14. Phrenic nerve
15. Scalenus medius muscle
16. Trapezius muscle
17. Omohyoid muscle
18. Clavicle
19. Superior trunk of brachial plexus
20. Posterior trunk of brachial plexus
21. Inferior trunk of brachial plexus
22. Left subclavian artery
23. Scalenus anterior muscle
24. Thoracic duct
25. Left internal jugular vein
26. External jugular vein
27. Left common carotid artery
28. Trachea
29. Inferior thyroid vein
30. Recurrent laryngeal nerve
31. Inferior thyroid artery
32. Sympathetic trunk
33. Thyroid gland
34. Cricothyroid muscle
35. Thyroid cartilage
36. Superior laryngeal nerve
37. Superior thyroid artery
38. Thyrohyoid membrane
39. Hyoid bone

3.2.2 Horizontal Section A

The plane of section (Fig. 27) passes through the larynx at the level of the rima glottidis (Figs. 25 and 26). The larynx lies ventral (prevertebral) to the cervical vertebrae. The infrahyoid muscles lie ventral to the larynx; dorsally, the hypopharynx is visible. The carotid sheath is prominent between the larynx and sternocleidomastoid muscle. Due to the obliquity of this section the ventral wall of an intervertebral disc has been cut (Fig. 27a). The vertebral artery can be seen within the foramen transversarium. Dorsal to this foramen one sees the anterior ramus of a spinal nerve destined for the brachial plexus. Dorsal to the transverse process of the vertebra is the articular process. Within the vertebral canal is the spinal cord. Lateral to the spinous process lie the muscles of the neck and shoulder (levator scapulae and trapezius).

In the CT image the vessels appear white due to contrast medium. The thyroid and arytenoid cartilages and the vocal folds lie in this section through the larynx.

Fig. 27. *a* and *b*. Neck horizontal section A.

1. Constrictor pharyngis muscle
2. Omohyoid muscle
3. Arytenoideus transversus muscle
4. Infrahyoid muscles
5. Plica vestibularis
6. Arytenoid cartilage
7. Thyroid cartilage
8. Platysma
9. Anterior jugular vein
10. Hypopharynx with piriform recess
11. Sternocleidomastoid muscle
12. Common carotid artery
13. Internal jugular vein
14. Scalenus anterior muscle
15. Brachial plexus
16. Scalenus medius and posterior muscles
17. Inferior articular process
18. Trapezius muscle
19. Levator scapulae muscle
20. Neck musculature
21. Spinal cord
22. Nuchal ligament
23. Spinous process of cervical vertebra
24. Transverse process of cervical vertebra
25. Vertebral artery
26. External jugular vein
27. Cervical vertebra
28. Vagal nerve
29. Intervertebral disc

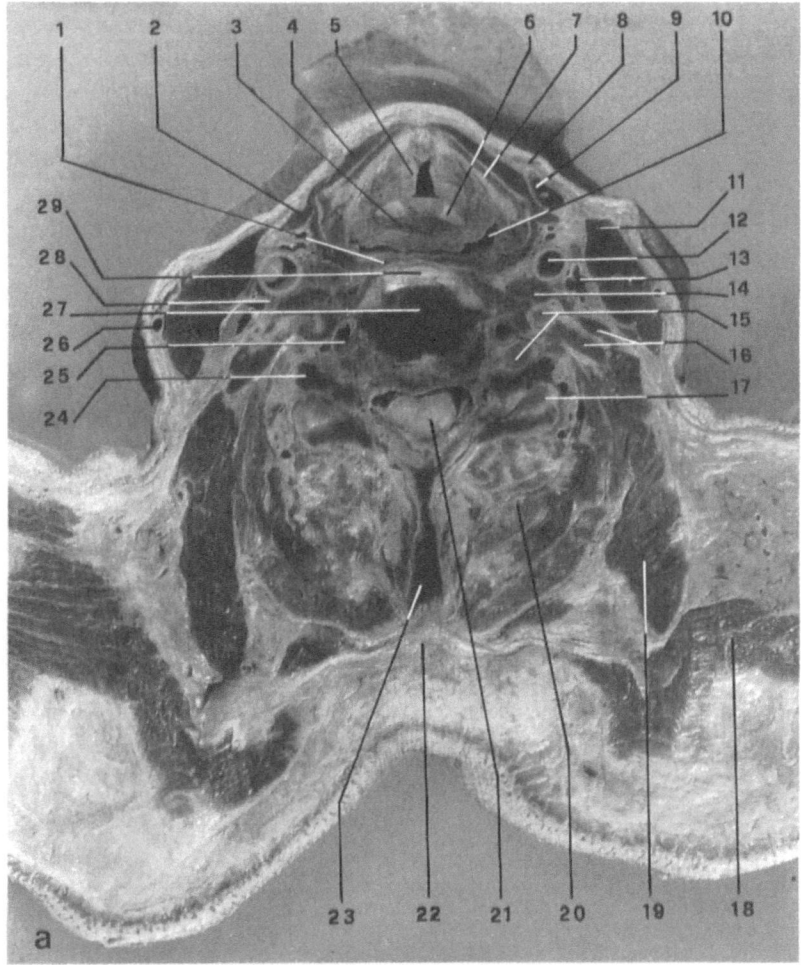

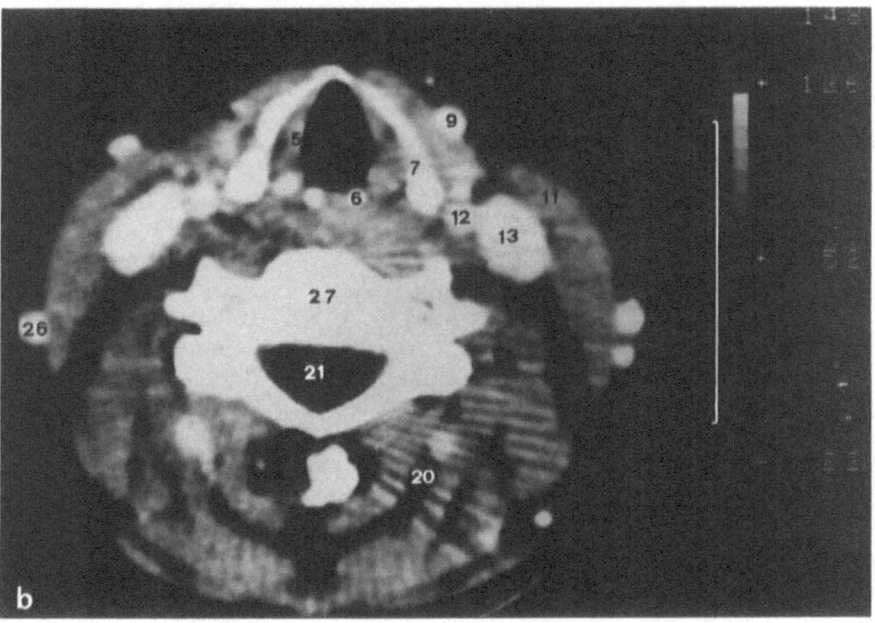

3.2.3 Horizontal Section B

This section transects the cricoid cartilage inferior to the rima glottidis (Fig. 28).

The topography of the musculature and vessels is similar to that of the previous section. Newly visible visceral structures are the cricoid cartilage, in-fraglottic portion of the larynx, and the esophagus (seen as a slit between the vertebral body and cricoid cartilage). On the right side, the superior pole of the thyroid gland and two dark, round lymph nodes are seen between the carotid artery and the cricoid cartilage. Due to oblique section two vertebral bodies and their intervertebral disc have been sectioned (Fig. 28a).

Fig. 28. *a* and *b*. Neck horizontal section *B*.

1. Sternocleidomastoid muscle	*16.* Scalenus medius and posterior muscles
2. Common carotid artery	
3. Thyroid gland	*17.* Articular process
4. Longus colli muscle	*18.* Neck musculature
5. Esophagus	*19.* Spinal cord
6. Cricoid cartilage	*20.* Spinous process, cervical vertebra
7. Cavum infraglotticum	*21.* Cervical vertebra (VII)
8. Infrahyoid muscles	*22.* Rhomboid minor muscle
9. Cricothyroid muscle	*23.* Levator scapulae muscle
10. Superior thyroid artery	*24.* Trapezius muscle
11. Inferior horn of thyroid cartilage	*25.* External jugular vein
12. Cervical vertebra (VI)	*26.* Brachial plexus
13. Vagal nerve	*27.* Vertebral artery
14. Scalenus anterior muscle	*28.* Internal jugular vein
15. Intervertebral disc	

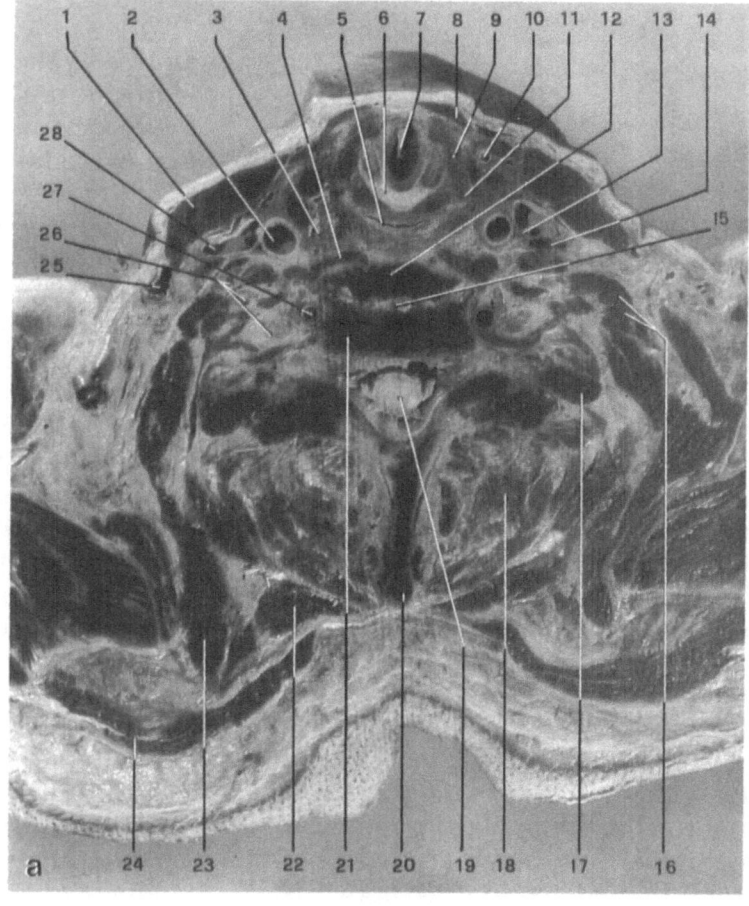

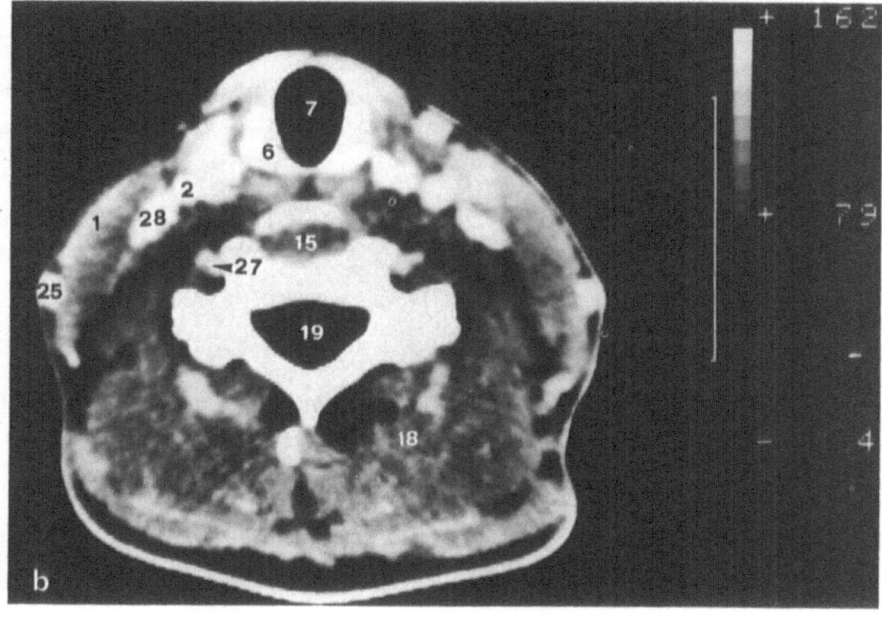

3.3. Thorax

3.3.1 Remarks

Figure 29 demonstrates the general organization of a transverse section through the thorax. The wall of the thorax consists of the sternum and costal cartilages ventrally, the ribs and intercostal muscles laterally, and the vertebral column and ribs dorsally.

The rib cage is surrounded externally by the pectoralis major and minor and serratus anterior muscles ventrally and muscles of the back dorsally. The scapula with its associated muscles is found above the seventh rib. In the female the mamma is located on the ventral surface of the thorax between the third and sixth rib.

The thoracic cavity consists of the two pleural cavities and the mediastinum. Anatomically the mediastinum is divided into four subdivisions. The ventral mediastinum is between the sternum and the pericardium and contains lymph nodes, adipose tissue, and thymus. The middle mediastinum is between the two lungs and contains pericardium, heart, ascending aorta, pulmonary trunk, lower half of the superior vena cava, and phrenic nerve. The dorsal mediastinum is between the pericardium and the vertebral column and contains esophagus, descending aorta, azygos vein, hemiazygos vein, thoracic duct, bifurcation of trachea, vagal nerve, and lymph nodes. The superior mediastinum lies superior to a line connecting the sternal angle to the superior surface of the fifth thoracic vertebra, and contains part of the thymus, the great vessels related to the heart including the aortic arch, the trachea, lymph nodes, and esophagus.

For purposes of clinical examination the thoracic cavity is divided into three tiers. The superior tier lies between the first and fourth thoracic vertebrae, the middle tier between the fourth and eighth thoracic vertebrae, and the inferior tier between the level of the valves and the centrum tendineum (Fig. 30). In this presentation we have chosen one section from each tier.

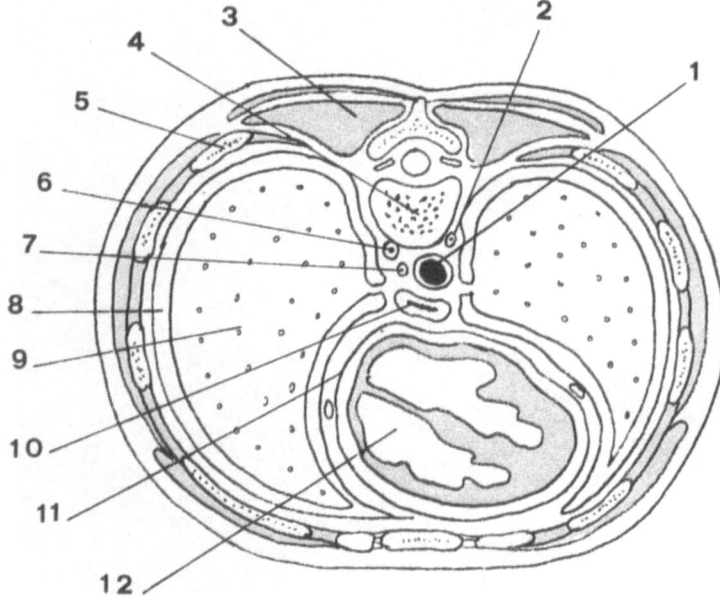

Fig. 29. Horizontal section of thorax.

1. Aorta, thoracic
2. Hemiazygos vein
3. Back musculature
4. Vertebral column
5. Ribs and intercostal muscles
6. Azygos vein
7. Thoracic duct
8. Pleural cavity
9. Lungs
10. Esophagus
11. Pericardium
12. Heart

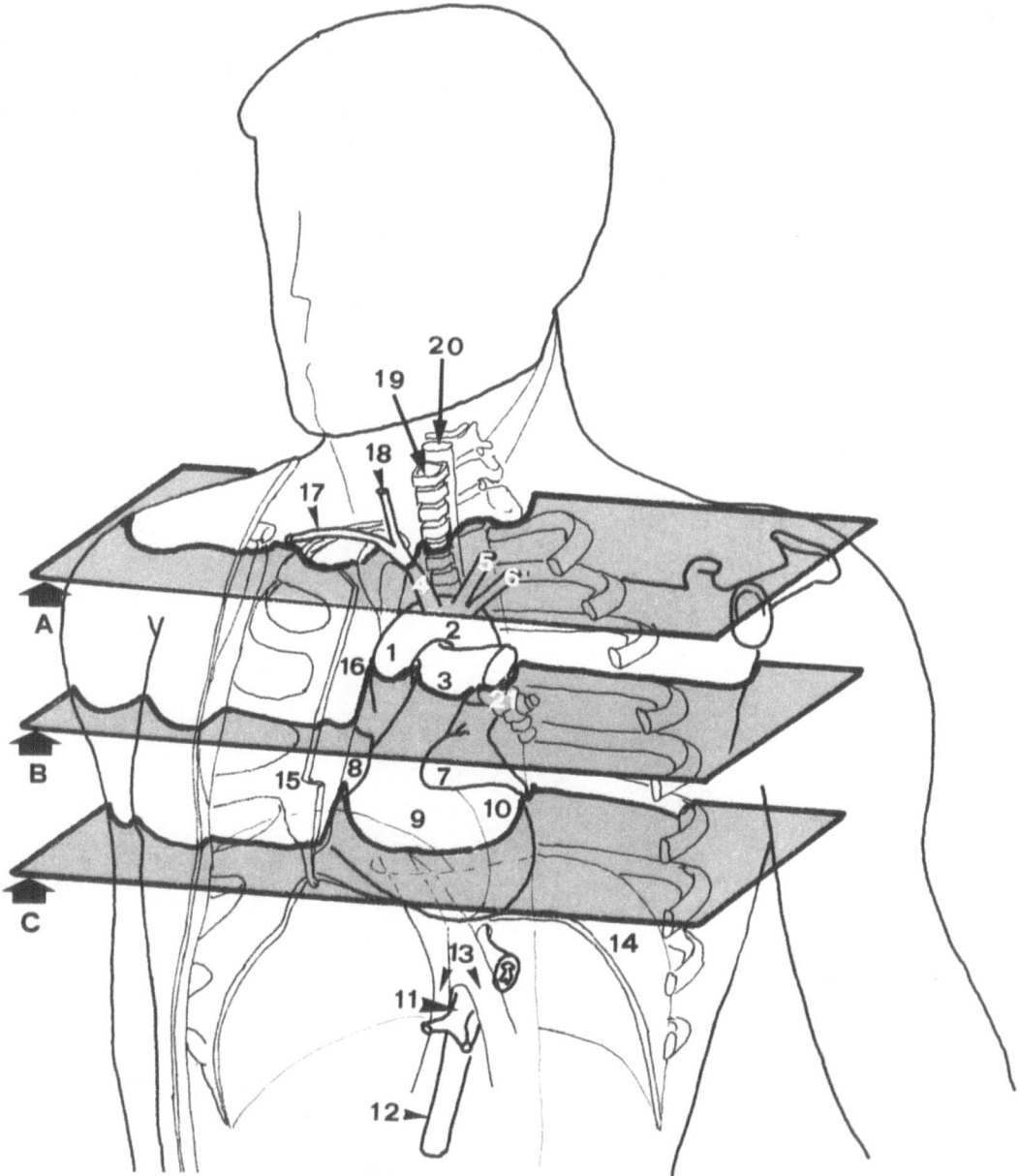

Fig. 30. Planes of section through thorax. *A*, upper level; *B*, middle level; *C*, lower level.

1. Aorta, ascending	*12.* Aorta, abdominal
2. Aortic arch	*13.* Right medial crus of diaphragm
3. Pulmonary trunk	*14.* Diaphragm
4. Brachiocephalic trunk	*15.* Sternum
5. Left common carotid artery	*16.* Lung
6. Left subclavian artery	*17.* Right subclavian artery
7. Left auricle	*18.* Left common carotid artery
8. Right auricle	*19.* Trachea
9. Right ventricle	*20.* Esophagus
10. Left ventricle	*21.* Left principal bronchus
11. Celiac trunk	

The structures of the mediastinum are presented from all sides (Fig. 31). Sections from the superior tier will transect the structures of the superior or supracardiac mediastinum. In the middle tier, in the hilus region, the pulmonary veins lie ventral, the pulmonary artery is in the middle, and the principal bronchus dorsal.

Notice on the *right* side: the arch of the azygos vein lies superior to the lobar bronchus to the superior lobe. The pulmonary arch lies between the lobar bronchus to the superior lobe and the lobar bronchus to the middle lobe. The aortic and pulmonary arches override the left principal bronchus.

Fig. 31. Planes of section through the mediastinum. *a.* Anterior aspect. *b.* Posterior aspect. *c.* Aspect from right. *d.* Aspect from left

1. Right brachiocephalic vein
2. Brachiocephalic trunk
3. Left brachiocephalic vein
4. Superior vena cava
5. Ascending aorta
6. Pulmonary trunk
7. Right pulmonary artery
8. Left pulmonary veins
9. Left atrium
10. Right atrium
11. Right ventricle
12. Thoracic aorta
13. Esophagus
14. Azygos vein
15. Bifurcation of trachea
16. Left common carotid artery
17. Right subclavian artery
18. Right auricle
19. Aortic arch
20. Left pulmonary artery
21. Left common carotid artery
22. Left subclavian artery
23. Left pulmonary veins
24. Left ventricle
25. Aortic valve
26. Right atrioventricular orifice
27. Left atrioventricular orifice
28. Inferior vena cava
29. Inferior thyroid artery
30. Vertebral artery
31. Inferior thyroid veins

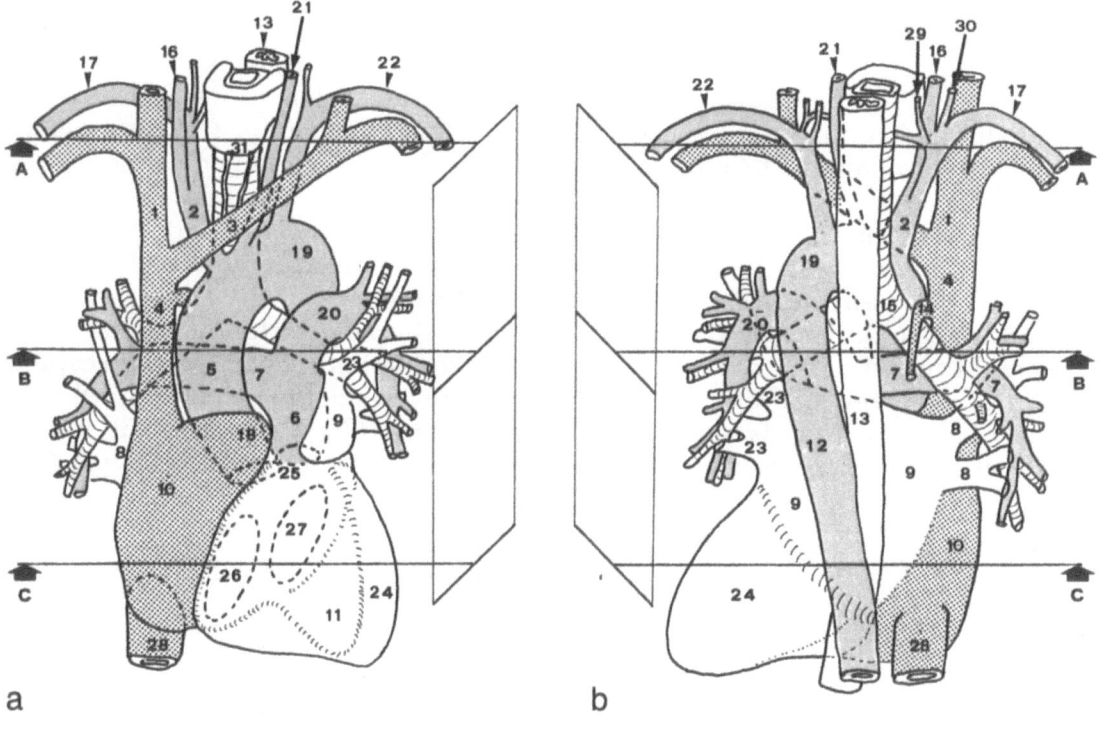

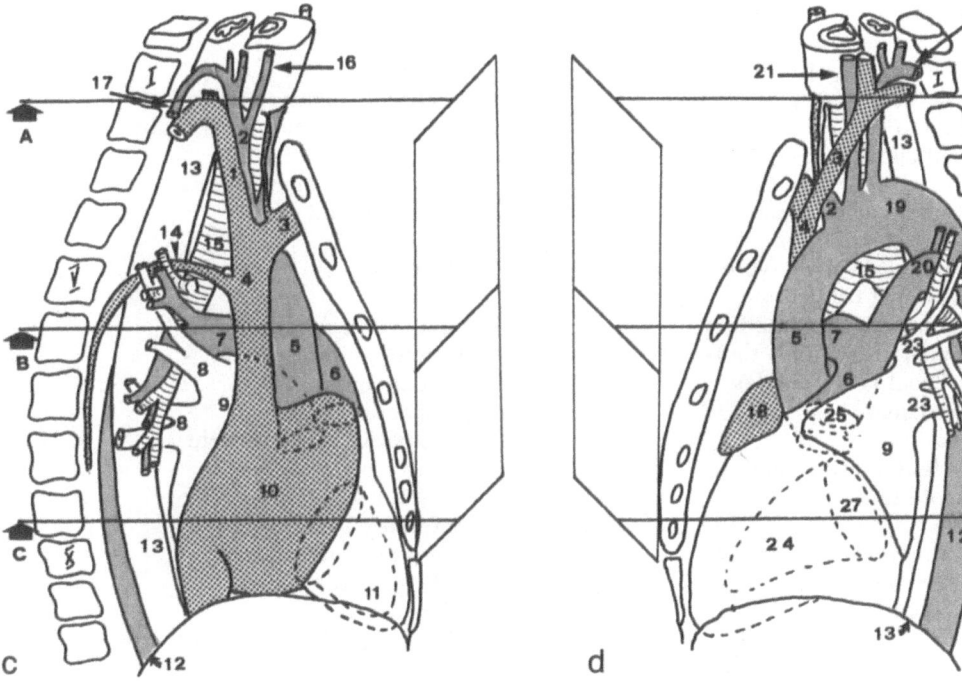

3.3.2 Horizontal Section A: Superior Tier

The central location of the trachea and esophagus is common to all sections through this tier (Fig. 32). The thyroid is sectioned at the level of the isthmus and shows an enlarged right lobe, part of which lies retrosternal (Fig. 32a). The arterial trunks lie lateral to the trachea and thyroid.

On the right the wide thoracic part of the subclavian artery is transected at the level of exit of the vertebral artery and thyrocervical trunk. Ventrolateral to the subclavian artery are the large venous trunks. The thoracic duct is cut at the level of its arch. The duct passes obliquely in a ventral direction from the tracheoesophageal angle between the left common carotid artery and the left subclavian artery to the venous angle. The apices of the lungs, sections of ribs, thoracic vertebral column with enclosed spinal cord, and the scapula with muscles of the shoulder girdle are well seen. The vasculature is not as well demonstrated in the CT image.

Fig. 32. *a* and *b*. Thorax: horizontal section A.

1. Right axillary artery	22. Esophagus
2. Cephalic vein	23. Pectoralis major muscle
3. Vertebral artery	24. Left superior pulmonary lobe
4. Right subclavian artery (pars cervicalis)	25. Second rib
5. Right subclavian vein	26. Second thoracic vertebra
6. Right subclavian artery	27. Intervertebral disc
7. Right brachiocephalic vein	28. Third rib
8. Thyrocervical trunk	29. Third thoracic vertebra
9. Right common carotid artery	30. Scapula
10. Subclavian artery (pars thoracalis)	31. Transverse process
11. Paratracheal lymph nodes	32. Erector spinae muscle
12. Thyroid gland	33. Spinal cord
13. Trachea	34. Trapezius muscle
14. Sternocleidomastoid muscle	35. Rhomboid major muscle
15. Infrahyoid muscles	36. Third rib
16. Clavicle	37. Serratus anterior muscle
17. Left common carotid artery	38. Intercostal muscles
18. Venous angle	39. Subscapularis muscle
19. Thoracic duct	40. Infraspinatus muscle
20. First rib	41. Second rib
21. Left subclavian artery	42. Right superior pulmonary lobe

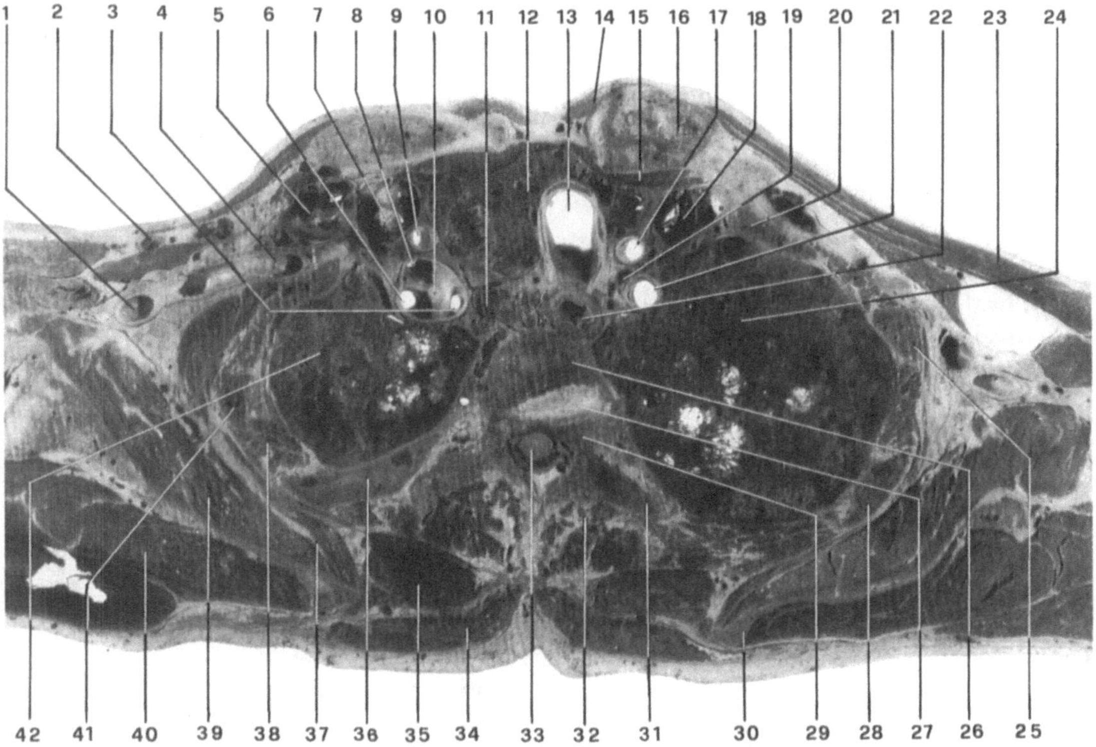

1 2 3 4 5 6 7 8 9 10 11 12 13 14 15 16 17 18 19 20 21 22 23 24
42 41 40 39 38 37 36 35 34 33 32 31 30 29 28 27 26 25

a

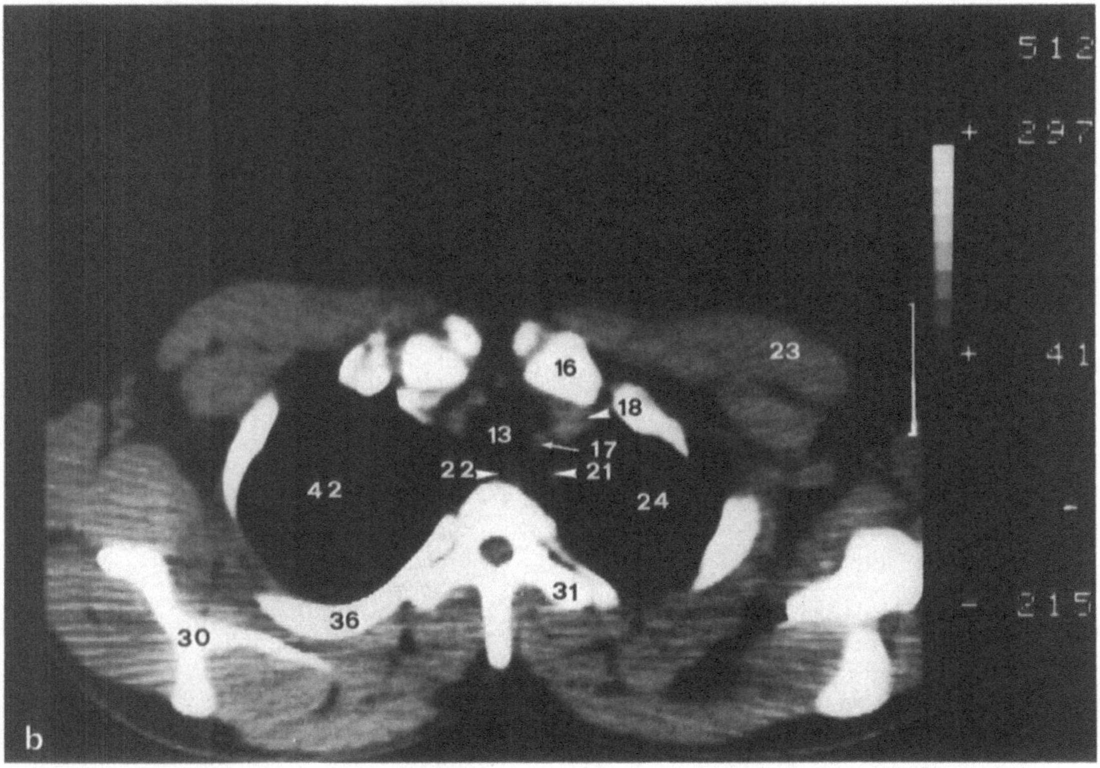

512
+ 297
+ 41
-
- 215
16
23
18
13
17
22
21
42
24
31
36
30

b

3.3.3 Horizontal Section *B*: Middle Tier

This section passes through the middle tier caudal to the bifurcation of the trachea (Fig. 33). The section shows the pulmonary trunk, the two pulmonary arteries, the ascending aorta, the principal bronchi, and laterally the hili of the lungs. In the dorsal medias-tinum the esophagus, descending thoracic aorta, azygos vein, and thoracic duct are visible. Notice that the esophagus lies on the left side at this level. The lung shows emphysematous changes on the left side.

The CT image was taken with a mediastinal window. Structures of the mediastinum are well defined, those of the lung are invisible.

Fig. 33. *a* and *b*. Thorax: horizontal section *B*.

1.	Rib	*17.*	Latissimus dorsi muscle
2.	Serratus anterior muscle	*18.*	Inferior pulmonary lobe
3.	Pectoralis major muscle	*19.*	Inferior bronchus (segment 6)
4.	Tracheobronchial lymph nodes	*20.*	Aorta, descending
5.	Right principal bronchus	*21.*	Esophagus
6.	Superior vena cava	*22.*	Thoracic duct
7.	Internal thoracic vessels	*23.*	Azygos vein
8.	Right pulmonary artery	*24.*	Thoracic vertebra (VI)
9.	Aorta, ascending	*25.*	Spinal cord
10.	Sternum	*26.*	Trapezius muscle
11.	Anterior mediastinal lymph node	*27.*	Pleural cavity
12.	Left pulmonary artery	*28.*	Inferior lobe of lung
13.	Costal cartilage	*29.*	Posterior mediastinal lymph nodes
14.	Left principal bronchus		
15.	Superior bronchus	*30.*	Oblique fissure
16.	Superior pulmonary lobe	*31.*	Inferior angle of scapula

1
2
3
4
5
6
7
8
9
10
11
12
13
14
15
16
31
30
29
28
27
26
25
24
23
22
21
20
19
18
17

a

10
12
16
6
9
1
8
21
20
24
31
28
18

b

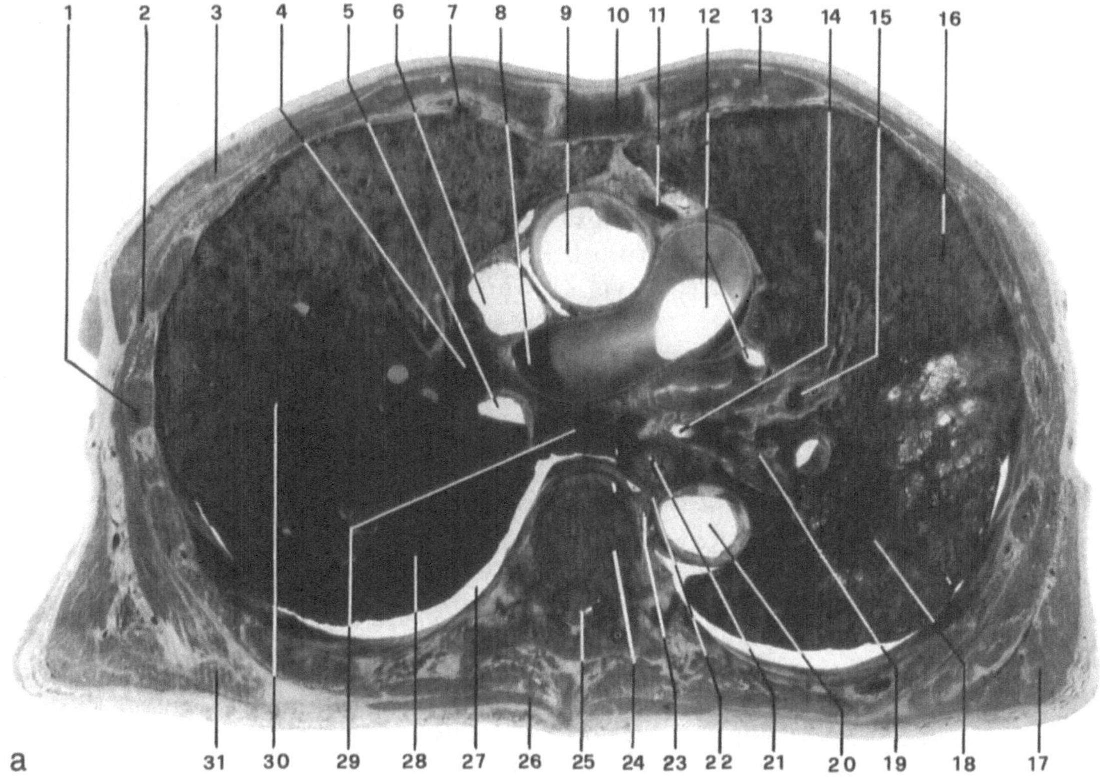

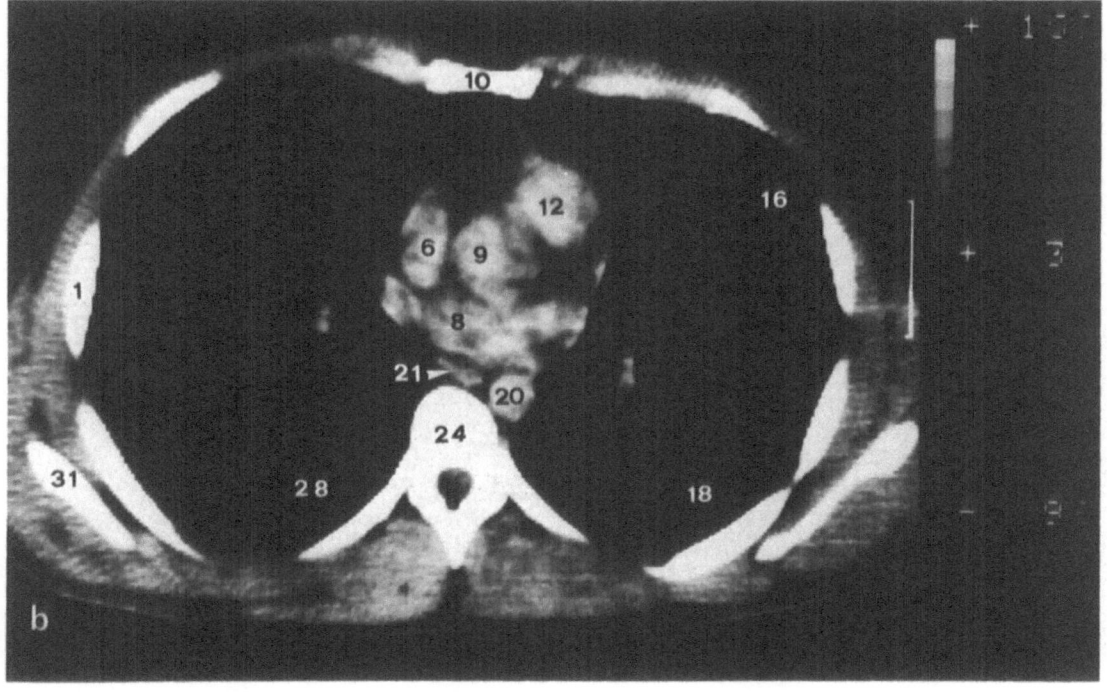

3.3.4 Horizontal Section *C*: Inferior Tier

Structures visible in the middle mediastinum are the two ventricles and two atria, the bicuspid and tricuspid valves, and the coronary arteries (Figs. 34 and 35). In the dorsal mediastinum one sees the thoracic aorta, azygos veins, thoracic duct, and esophagus. Note that the esophagus lies on the left side. In the region of the right pleural cavity one sees the apex of the diaphragm and the underlying liver (Fig. 34, plane of section *C*). The left lower lobe and ventral aspect of the right lower lobe demonstrate emphysematous changes.

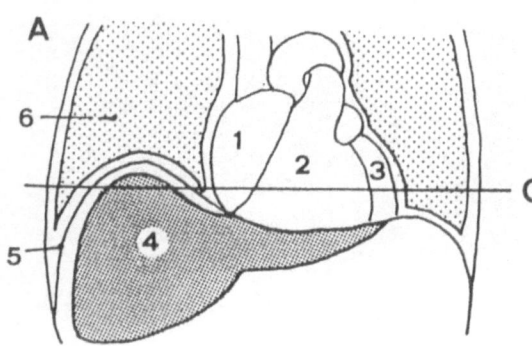

Fig. 34. Thorax: horizontal section *C*.

C Plane of section
1. Right atrium
2. Right ventricle
3. Left ventricle
4. Liver
5. Diaphragm
6. Right lung

Fig. 35. *a* and *b*. Thorax: horizontal section *C*.

1. Oblique fissure
2. Rib
3. Right atrium
4. Left atrium
5. Medial lobe of lung
6. Interatrial septum
7. Left atrioventricular valve
8. Right atrioventricular valve
9. Right coronary artery
10. Sternum
11. Costal cartilage
12. Right ventricle
13. Interventricular septum
14. Anterior papillary muscle
15. Anterior interventricular ramus of left coronary artery
16. Trabeculae carneae
17. Lingula pulmonis
18. Left ventricle
19. Chordae tendineae
20. Left marginal artery
21. Circumflex branch of coronary artery
22. Pericardial cavity
23. Coronary sinus
24. Inferior lobe of lung
25. Aorta, descending
26. Esophagus
27. Hemiazygos vein
28. Thoracic duct
29. Spinal cord
30. Transversospinalis muscle
31. Thoracic vertebra
32. Erector spinae muscle
33. Pleural cavity
34. Inferior lobe of lung
35. Diaphragm
36. Liver
37. Lattissimus dorsi muscle

1
2
3
4
5
6
7
8
9
10
11
12
13
14
15
16
17
18
19
20
21
22
23
24
25
26
27
28
29
30
31
32
33
34
35
36
37

a

5
12
17
18
2
36
26
25
31
34
24

b

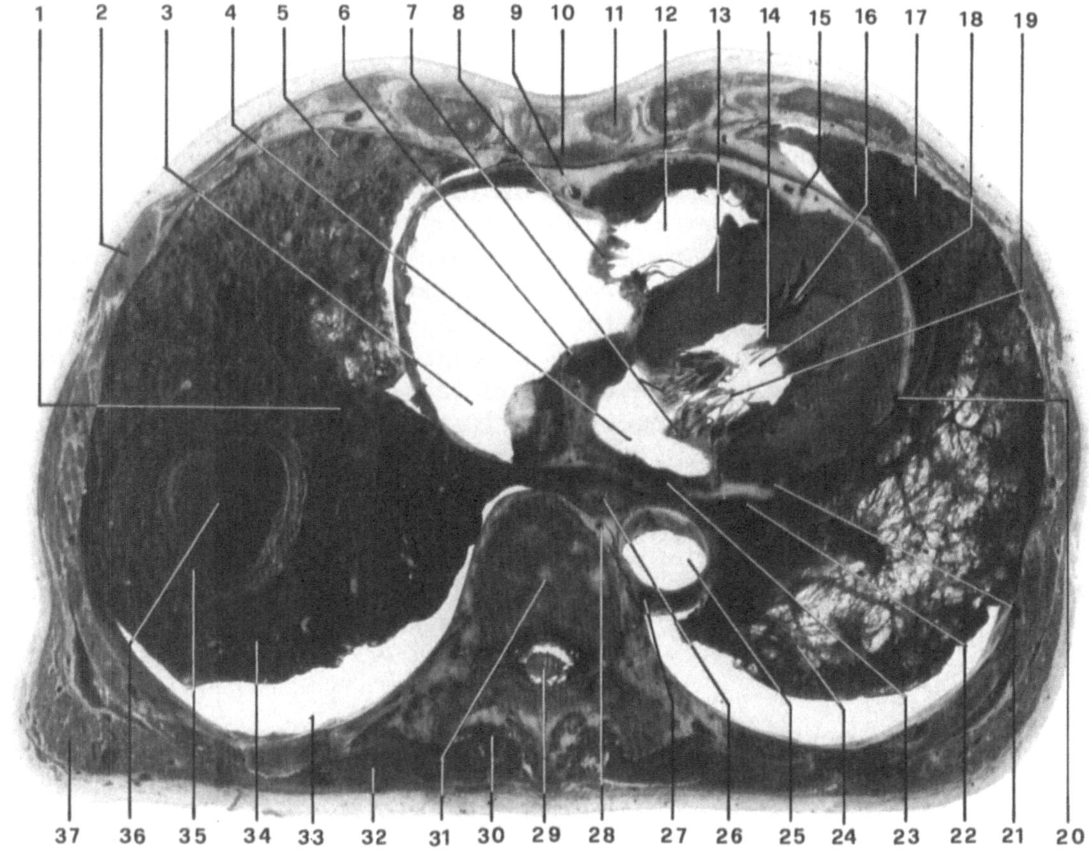

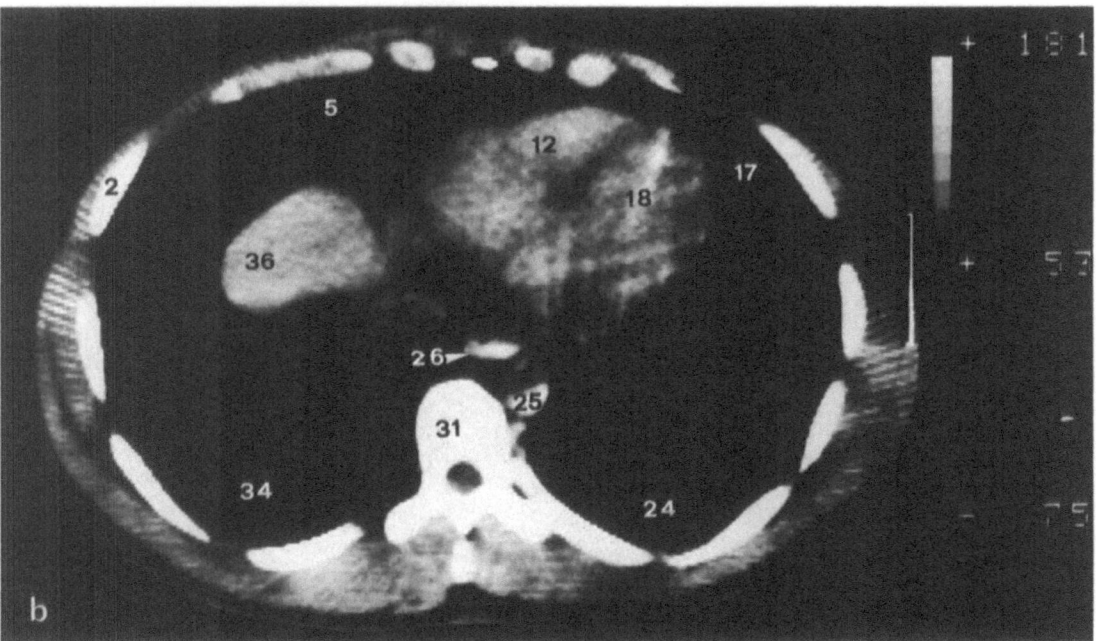

3.4 Abdomen

The abdominal cavity extends from the diaphragm to the plane of the superior aperture of the lesser pelvis (linea terminalis). The transverse colon divides the abdomen into a supracolic and an infracolic division. The dorsal abdominal wall borders on the retroperitoneal space. The supracolic division contains the liver, the stomach, the superior portion and the superior third of the descending portion of the duodenum, the pancreas, and the spleen. At this level the kidneys and adrenal glands lie in the retroperitoneal space.

Tomograms are most frequently ordered for the supracolic region (Fig. 36).

The infracolic region contains the small intestine which is surrounded by the divisions of the colon.

Since the dome-shaped diaphragm extends into the thoracic cavity the ribs make up the lateral wall of the supracolic region. The recti abdominis muscles and portions of the oblique and transverse abdominal muscles attach anteriorly to the costal margins of the thoracic cage.

Figure 36 shows the three levels of section previously described, and aids in determining which abdominal organs will be found in a certain plane.

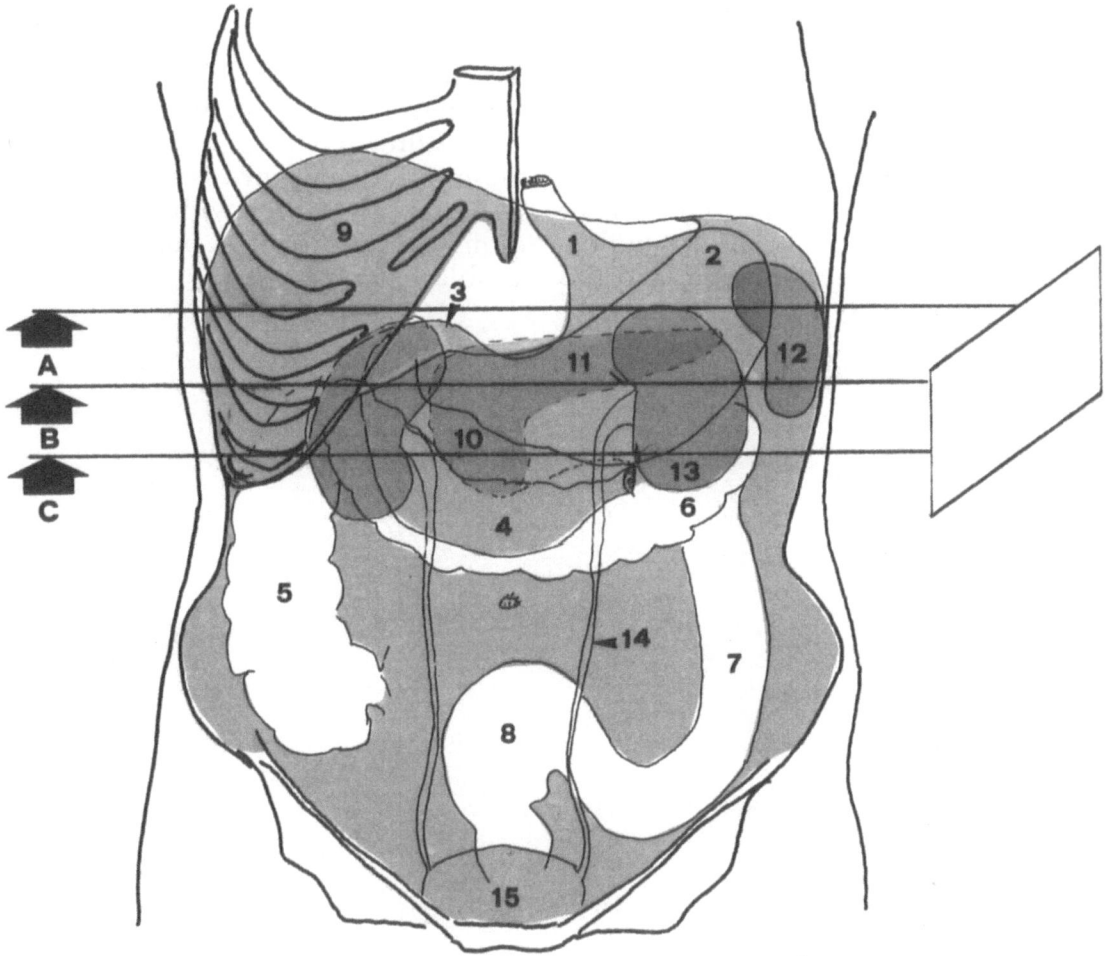

Fig. 36. Schematic projection of the organs of the abdominal cavity: levels of planes of section *A*, *B*, and *C*.

1. Cardia of stomach
2. Fundus of stomach
3. Bulb of duodenum
4. Pars transversa of duodenum
5. Ascending colon
6. Left colic flexure
7. Descending colon
8. Sigmoid colon
9. Liver
10. Head of pancreas
11. Body of pancreas
12. Spleen
13. Kidney
14. Ureter
15. Urinary bladder

3.4.1 Horizontal Section A

In this section the following structures have been transected: liver, inferior vena cava, aorta, fundus of stomach, spleen (pathologically enlarged), superior pole of left kidney, and lower margin of left lung (Fig. 37). A tumor is seen between the diaphragm and the pleura. The tail of the pancreas is seen in the CT image between the left kidney and the body of the stomach. The CT image is from a patient with a left-sided adrenal tumor, which displaces the inferior vena cava ventromedially. The stomach is filled with contrast medium, thus providing a prominent magenblase.

Fig. 37. *a* and *b*. Upper abdomen: horizontal section A.

1.	Rib	*15.*	Spleen
2.	Right lobe of liver	*16.*	Splenic artery and vein
3.	Hepatic veins	*17.*	Kidney
4.	Costal cartilage	*18.*	Renal sinus
5.	Inferior vena cava	*19.*	Suprarenal gland
6.	Caudate lobe of liver	*20.*	Thoracic duct
7.	Hepatogastric ligament	*21.*	Spinal cord
8.	Diaphragm, pars lumbalis	*22.*	Thoracic vertebra
9.	Left lobe of liver	*23.*	Aorta, abdominal
10.	Stomach with gastric plicae	*24.*	Azygos vein
11.	Tumor	*25.*	Latissimus dorsi muscle
12.	Diaphragm	*26.*	Subdiaphragmatic space
13.	Inferior lobe of lung	*27.*	Pancreas, tail
14.	Pleural cavity	*28.*	Tumor of suprarenal gland

1
2
3
4
5
6
7
8
9
10
11
12
13
26
25
24
23
22
21
20
19
18
17
16
15
14

a

9
1
2
5
10
28
23
27
22
15
21
17

b

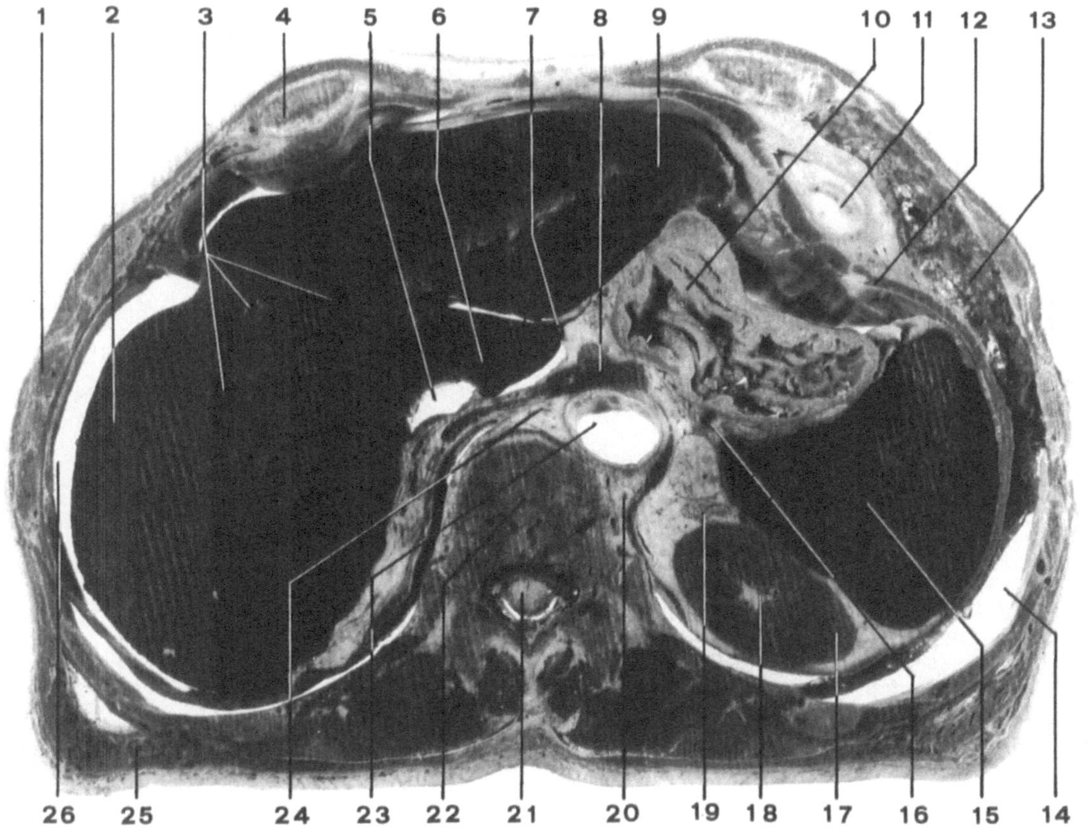

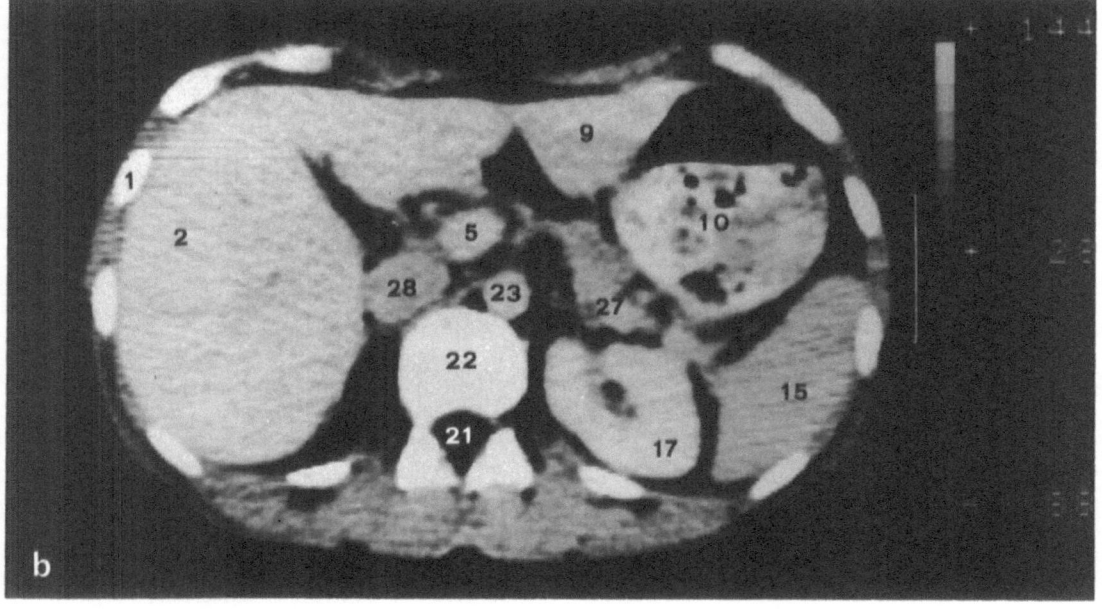

3.4.2 Horizontal Section B (Fig. 38)

This section transects within the peritoneal cavity the liver at the level of the portal fissure (portal vein, hepatic artery, hepatic ducts), inferior vena cava, stomach, and spleen; and within the retroperitoneal space the kidneys, suprarenal glands, tail of the pancreas, splenic artery, and abdominal aorta at the level of the celiac trunk (Fig. 38). The omental bursa appears as a slit between the dorsal stomach wall and the ventral surface of the pancreas.

In the CT image one sees liver, stomach, pancreas (pathologically enlarged), spleen, and gallbladder. On the right, as in the previous section A, a tumor of the suprarenal gland is visible.

Fig. 38. *a* and *b*. Upper abdomen: horizontal section *B*.

1.	Rib	*18.*	Spleen
2.	Hepatic vein	*19.*	Left suprarenal gland
3.	Right hepatic artery	*20.*	Left renal vein
4.	Hepatic duct	*21.*	Left renal artery
5.	Portal vein	*22.*	Left suprarenal vein
6.	Left hepatic artery	*23.*	Ascending lumbar vein
7.	Falciform ligament	*24.*	Thoracic duct
8.	Inferior vena cava	*25.*	Spinal cord
9.	Left lobe of liver	*26.*	Sacrospinal muscle
10.	Proper hepatic artery	*27.*	Quadratus lumborum muscle
11.	Celiac trunk	*28.*	Psoas major muscle
12.	Splenic artery	*29.*	Kidney
13.	Stomach	*30.*	Fat capsule
14.	Peritoneal cavity	*31.*	Right lobe of liver
15.	Abdominal aorta	*32.*	Tumor of suprarenal gland
16.	Tail of pancreas	*33.*	Gallbladder
17.	Pleural cavity		

1
2
3
4
5
6
7
8
9
10
11
12
13
14
15
16
17
18
19
20
21
22
23
24
25
26
27
28
29
30
31

a

33
13
32
15
16
31
18
29
29

b

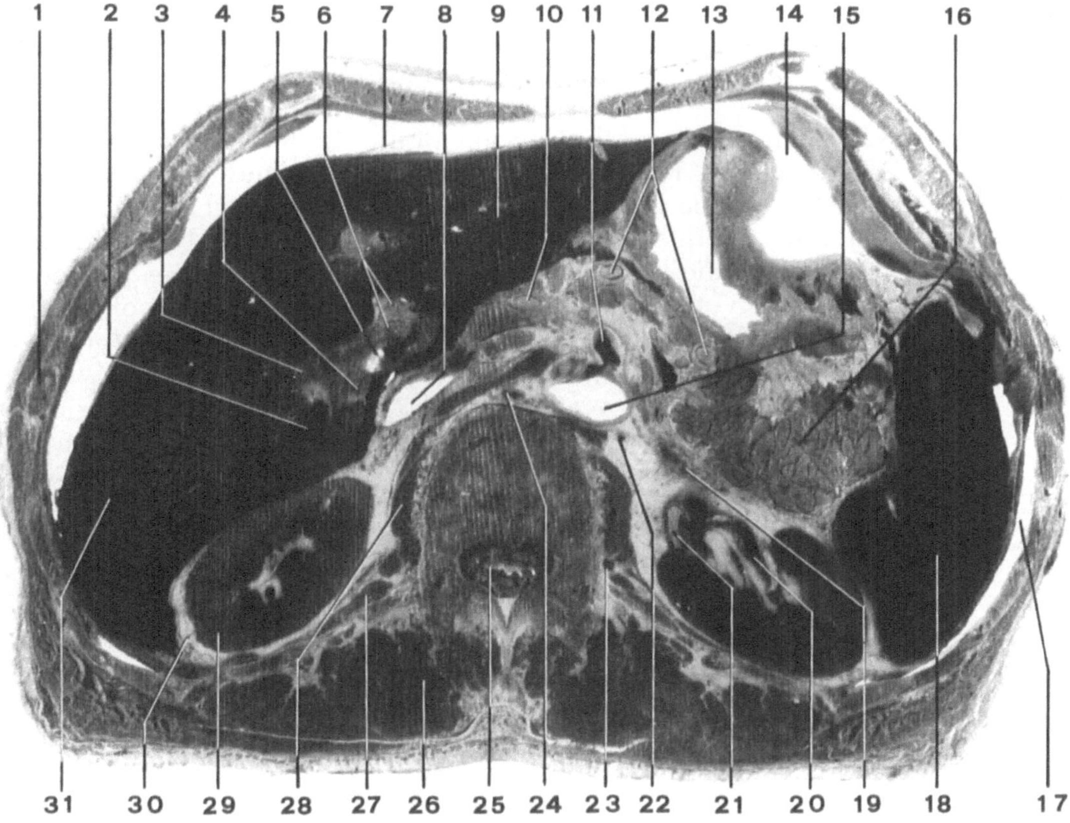

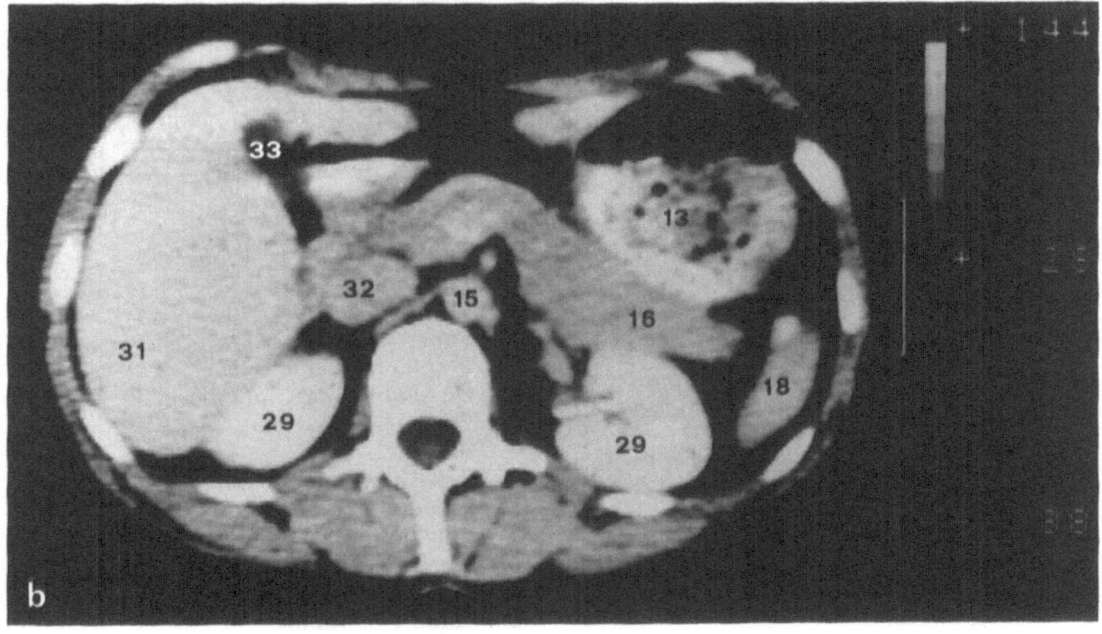

3.4.3 Horizontal Section *C*

The plane of section transects all three divisions of the pancreas, i.e., head, body, and tail (Fig. 39). The principal organs and structures in this section are the liver with part of the portal fissure, pancreas, inferior vena cava, superior mesenteric vein, splenic vein, right renal vein, aorta, right and left renal arteries, two kidneys, spleen, left colic flexure, and pyloric antrum.

The CT image shows the superior pole of the right kidney and a tumor of the suprarenal gland lying ventral to it. The left suprarenal gland is seen as a star-shaped structure lying lateral to the abdominal arota. The pyloric antrum and undefined loops of small intestine are visible in the ventral portion of the abdominal cavity.

Fig. 39. *a* and *b*. Upper abdomen: horizontal section *C*.

1.	Right renal artery	17.	Kidney
2.	Head of pancreas	18.	Renal pelvis
3.	Left and right hepatic ducts	19.	Left renal artery
4.	Inferior vena cava	20.	Left renal vein
5.	Gastroduodenal artery	21.	Thoracic vertebra
6.	Superior mesenteric vein	22.	Spinal cord
7.	Body of pancreas	23.	Erector spinae muscle
8.	Splenic vein	24.	Quadratus lumborum muscle
9.	Omental bursa	25.	Psoas major muscle
10.	Superior mesenteric artery	26.	Right renal vein
11.	Antrum of stomach	27.	Latissimus dorsi muscle
12.	Abdominal aorta	28.	Liver
13.	Left colic flexure	29.	Left suprarenal gland
14.	Gastrocolic ligament	30.	Small intestine
15.	Spleen	31.	Superior pole of kidney
16.	Splenic vein	32.	Suprarenal tumor

1
2
3
4
5
6
7
8
9
10
11
12
13
14
15
16
17
18
19
20
21
22
23
24
25
26
27
28

a

11
30
29
4
12
32
28
31
17
+ 234
+ 65
- 96

b

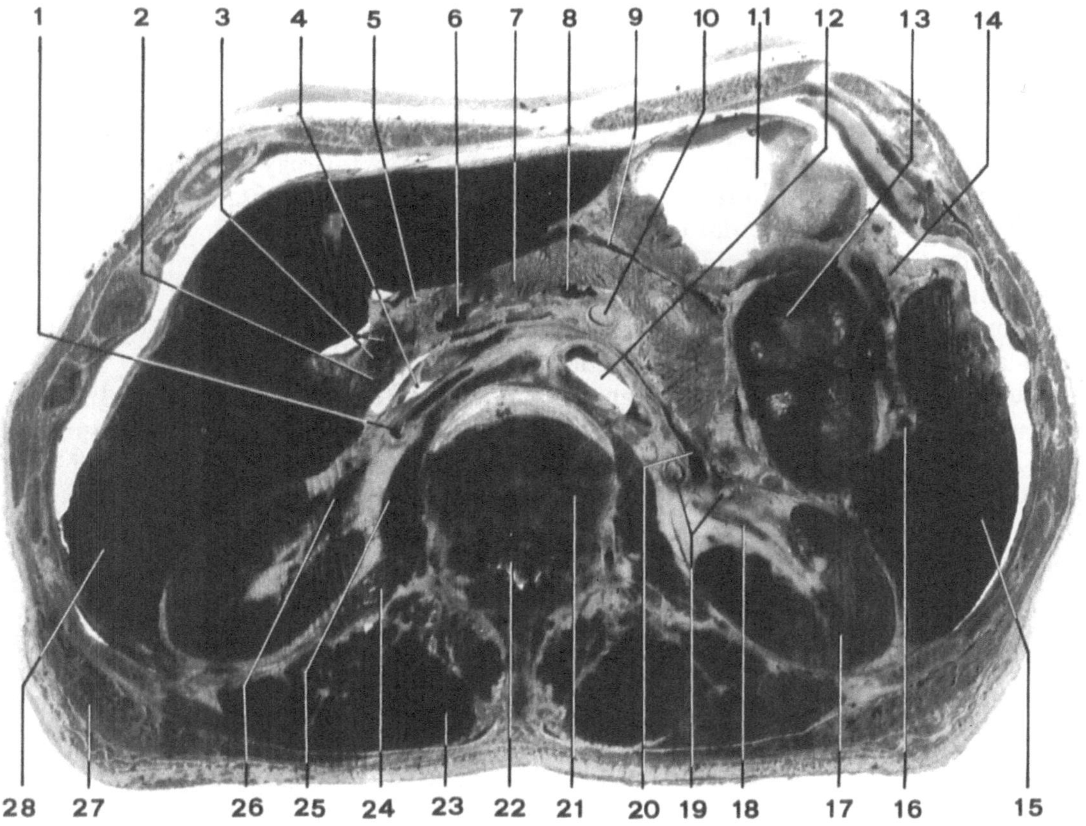

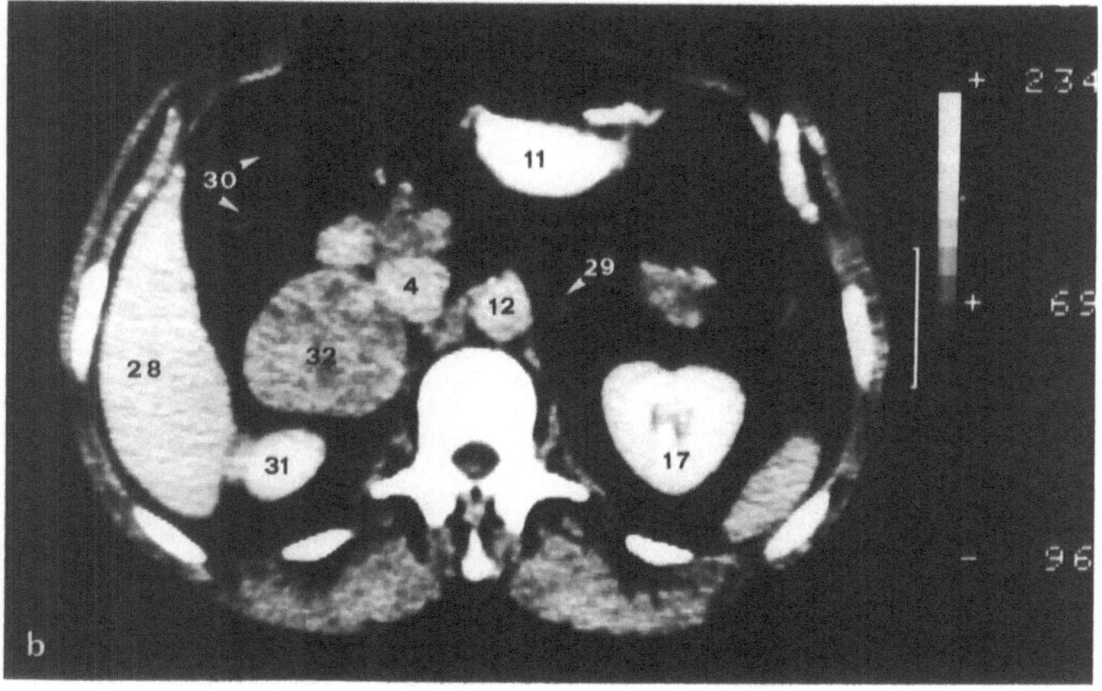

3.5 Pelvis and Hip

3.5.1 Male Pelvis

The walls of the pelvis and the adjacent hip are similar in both sexes (Fig. 40). The bones and articulations of the pelvis consist of the sacrum, innominate bone (ilium, pubis, and ischium), proximal femur, symphysis, and sacroiliac and hip joints. On the external surface of the pelvic girdle appear the gluteal muscles and lateral rotators, on the internal surface the internal obturator and the levator funnel, and caudally the ischiorectal fossa. Branches of the internal iliac artery lie on the dorsolateral pelvic wall.

The male pelvic cavity contains the rectum posteriorly, the bladder ventrally, and the prostate inferior to the bladder. The seminal vesicles lie on the dorsal surface of the bladder, flanked by the terminal portions of the ureters and ductus deferentes (Fig. 40).

In this region the rectum, bladder, and prostate are examined. Figure 40 shows the planes of section and their relationships to the various organs.

Fig. 40. Male pelvis. *a.* Frontal section. *b.* Lateral aspect.

A–C Planes of section
1. Abdominal aorta
2. Right common iliac artery
3. Right internal iliac artery
4. Right external iliac artery
5. Inferior mesenteric artery
6. Inferior vena cava
7. Right internal iliac vein
8. Right external iliac vein
9. Left ureter
10. Psoas major muscle
11. Iliacus muscle
12. Internal obturator muscle
13. Inferior ramus of pubic bone
14. Urogenital diaphragm
15. Levator ani muscle
16. Acetabulum
17. Head of femur
18. Hip joint
19. Ductus deferens
20. Obturator artery
21. Obturator nerve
22. Urinary bladder
23. Urethra
24. Prostate gland
25. Rectum

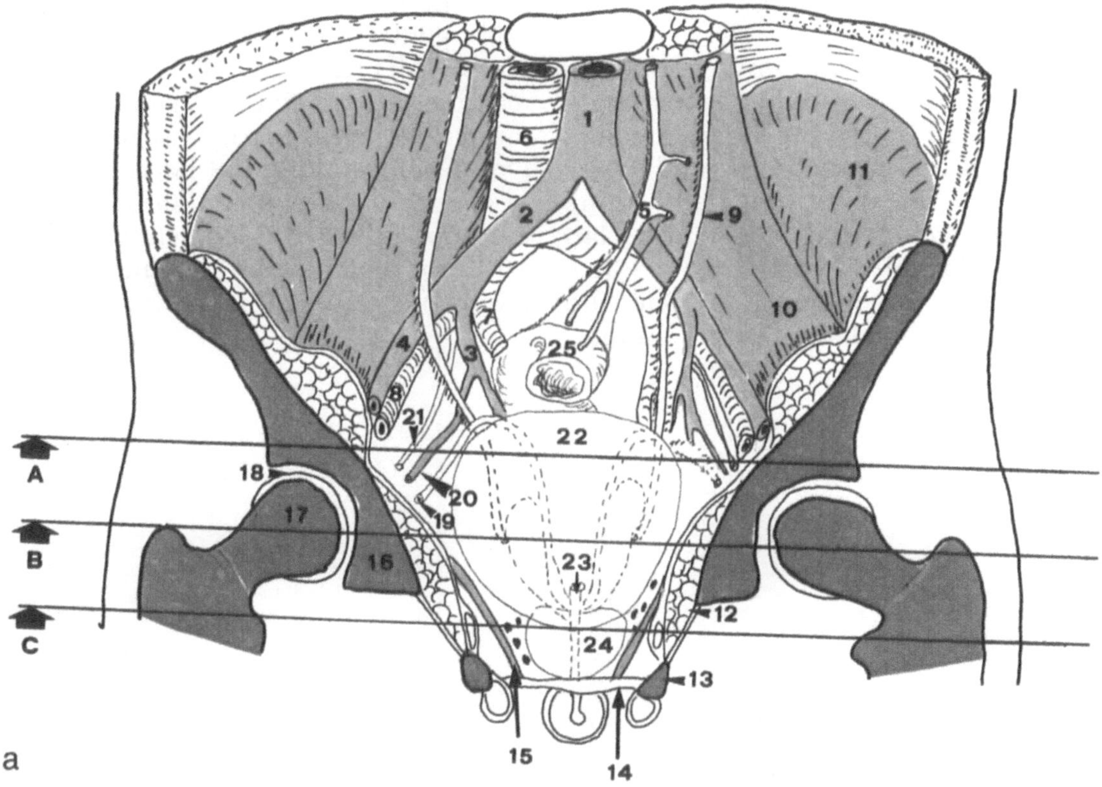

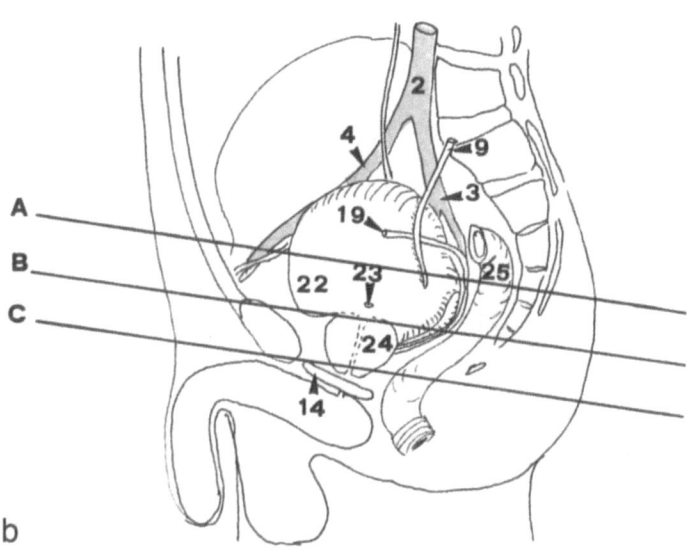

a

b

3.5.1.1 Horizontal Section A

The bladder is in the center of the section, ventral to which lie several convolutions of ileum, the sigmoid colon, and the external iliac vein and artery (Fig. 41). Dorsal to the bladder one sees the rectum and sacrum. The lateral pelvic wall is formed by the iliac bone. The three gluteal muscles lie lateral and dorsolateral to the ilium.

In the CT image the bladder, filled with contrast medium, almost completely fills the pelvic cavity. Prior to making the CT image lymphangiography was performed; as a result the lymph nodes along the lateral margin of the pelvis are filled with contrast medium and well demonstrated. Dorsal to the bladder the rectum and sigmoid colon are visible.

Fig. 41. *a* and *b*. Male pelvis: horizontal section A.

1.	Ilium (Os)	13.	Gluteus medius muscle
2.	Tensor fasciae latae muscle	14.	Gluteus maximus muscle
3.	Sartorius muscle	15.	Sciatic nerve
4.	Femoral nerve	16.	Inferior gluteal vessels
5.	Iliopsoas muscle	17.	Rectal vessels
6.	Convolutions of ilium	18.	Rectum
7.	Rectus abdominis muscle	19.	Sacrum
8.	Mesentery	20.	Urinary bladder
9.	Sigmoid colon	21.	Ureter
10.	Left external iliac artery	22.	Internal obturator muscle
11.	Left external iliac vein	23.	External iliac lymph nodes
12.	Gluteus minimus muscle		

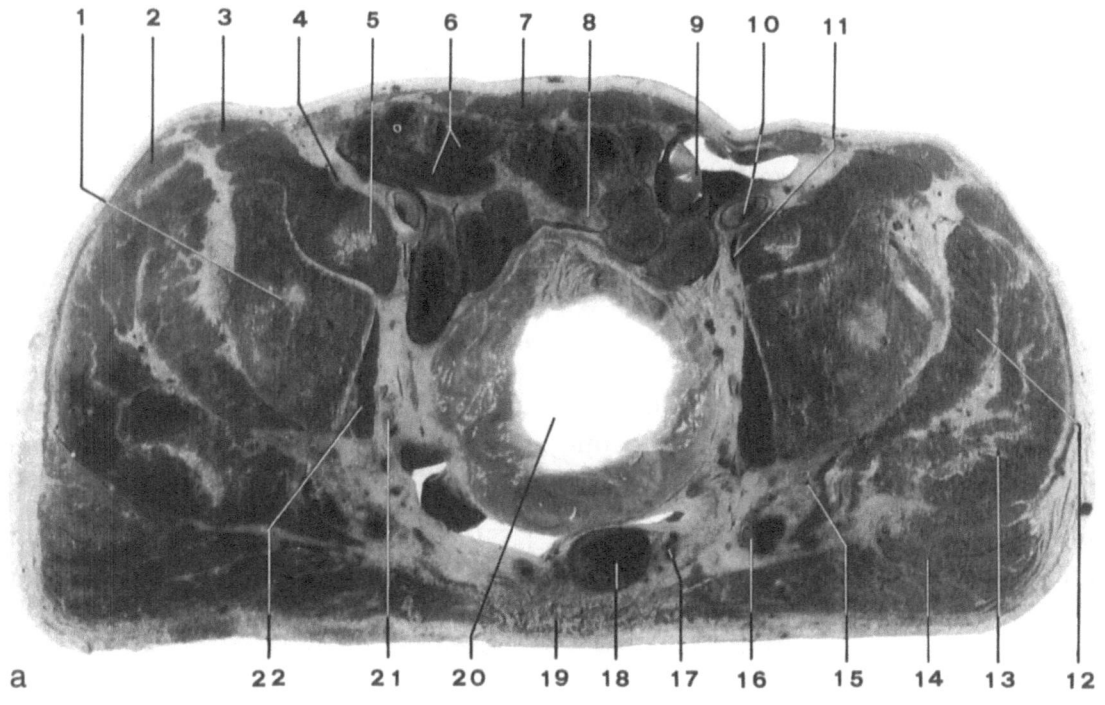

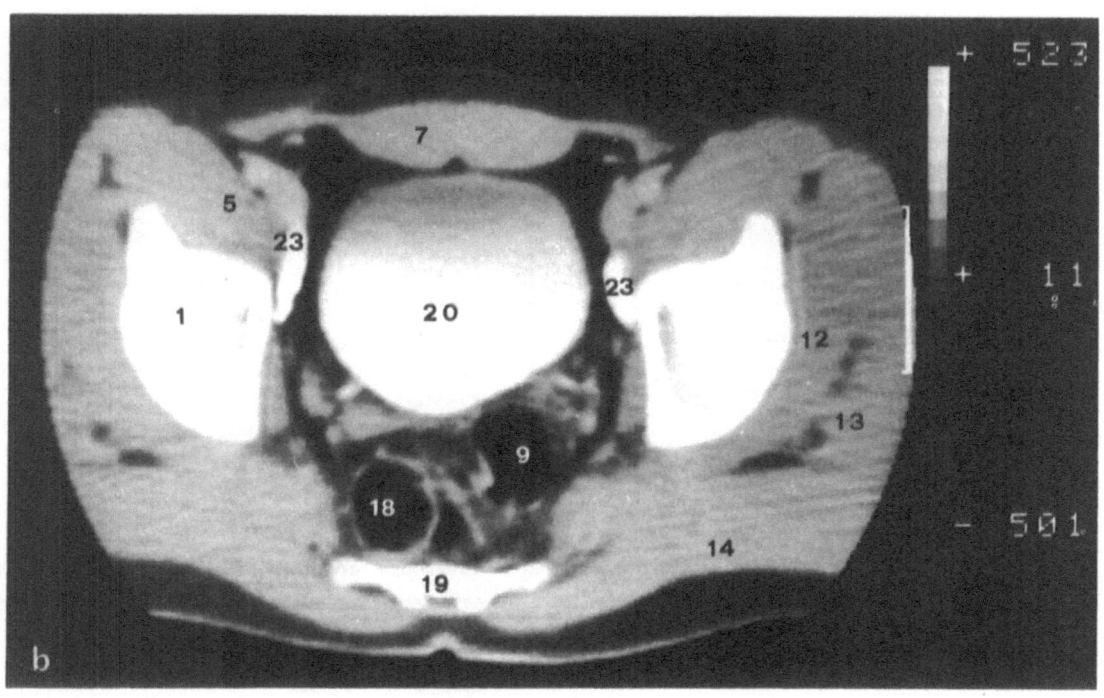

3.5.1.2 Horizontal Section B

The plane of section is through the symphysis and the femoral heads (Fig. 42). The internal obturator muscles and levator ani muscles are seen dorsal to the pubic bone and medial to the ischium. The right internal obturator muscle is seen curving around the ischium and into the trochanteric fossa. Within the levator funnel (from ventral to dorsal) are the base of the bladder, the prostate with the urethra, the seminal vesicles, and the rectum. The lateral aspect of the section shows the region of the hip joint, lateral rotators, and gluteus maximus muscle. Between the internal obturator muscle and levator ani muscle is the fatty tissue of the ischiorectal fossa.

Note: The ischiorectal fat is a continuation of the subcutaneous fat. It passes between the gluteal muscles into the ischiorectal fossa (well demonstrated in the CT image). The spermatic cord lies just ventral to the pubic tubercle; the femoral vessels lie lateral to the spermatic cord in the iliopectineal fossa of the subinguinal region.

In the CT section the following structures are visible: bladder filled with contrast medium, rectum filled with air, ischiorectal fossa filled with fat, hip joint, and femoral vessels. The prostate and seminal vesicles are not visible.

Fig. 42. *a* and *b*. Male pelvis: horizontal section *B*.

1.	Fascia lata (tractus iliotibialis)	*16.*	Neck of femur
2.	Iliofemoral ligament	*17.*	Quadratus femoris muscle
3.	Tensor fasciae latae muscle	*18.*	Sciatic nerve
4.	Rectus femoris muscle	*19.*	Inferior gluteal vessels
5.	Sartorius muscle	*20.*	Prostate gland
6.	Iliopsoas muscle	*21.*	Urethra (pars urethrae)
7.	Obturator vessels	*22.*	Ductus deferens
8.	Pectineus muscle	*23.*	Rectum
9.	Symphysis	*24.*	Seminal vesicles
10.	Urinary bladder	*25.*	Levator ani muscle
11.	Spermatic cord	*26.*	Internal obturator muscle
12.	Pubic bone	*27.*	Ischium
13.	Common femoral vein	*28.*	Gluteus maximus muscle
14.	Common femoral artery	*29.*	Ischiorectal fossa
15.	Head of femur	*30.*	Subcutaneous fat

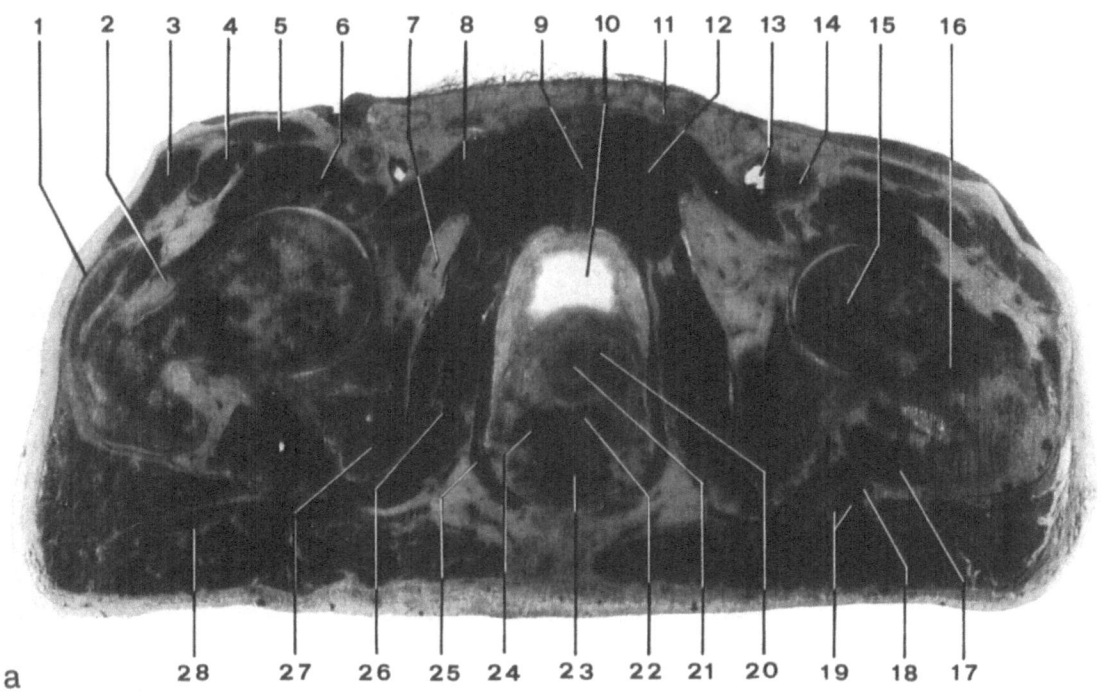

a

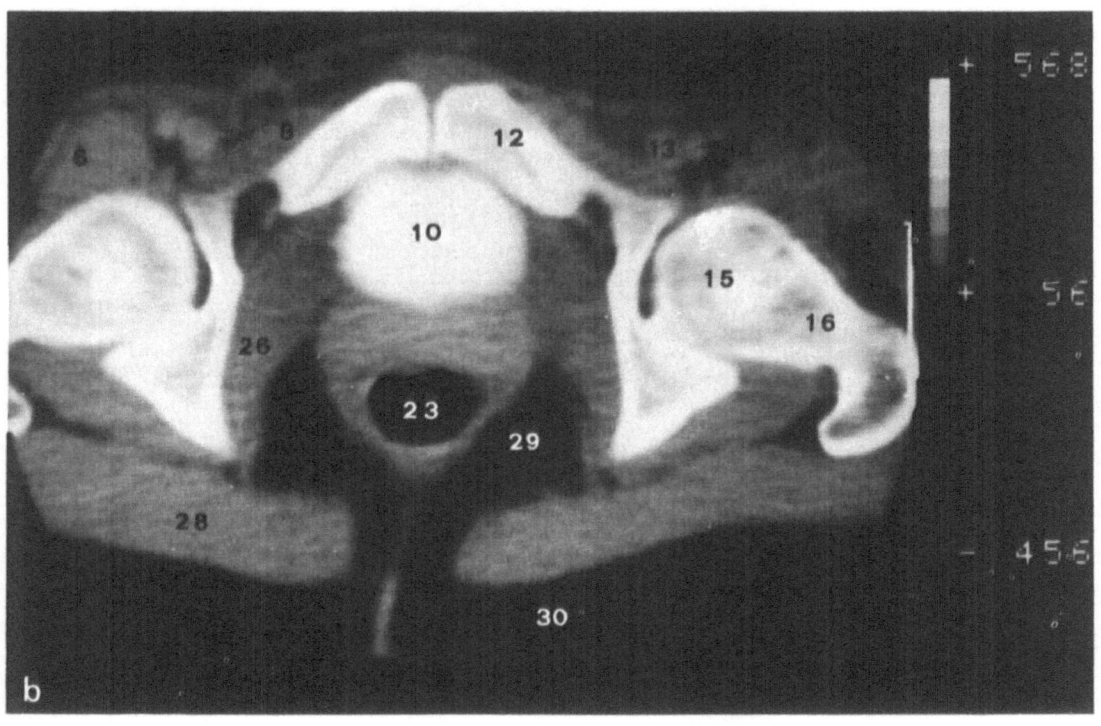

b

3.5.1.3 Horizontal Section C

The plane of section transects the ischial tuberosities and the inferior rami of the pubic bones (Fig. 43). In the midline region (from ventral to dorsal) note the inferior rami of the pubic bones within the levator funnel, the venous prostatic plexus, the prostate, and the rectum. The ischiorectal fossa lies between the internal obturator muscle and the levator ani muscle.

The fatty tissue of the ischiorectal fossa is a continuation of the subcutaneous fat. The hip joint is seen laterally; the quadratus femoris muscle and gluteus maximus muscle are visible in the gluteal region. The external obturator muscle is prominent between the inferior ramus of the pubic bone and the neck of the femur. The femoral vessels are well shown in the iliopectineal fossa.

In the CT image most of the above listed structures are visible.

Fig. 43. *a* and *b*. Male pelvis: horizontal section *C*.

 1. Fascia lata (tractus iliotibialis)
 2. Tensor fasciae latae muscle
 3. Rectus femoris muscle
 4. Sartorius muscle
 5. Iliopsoas muscle
 6. Obturator externus muscle
 7. Pectineus muscle
 8. Venous plexus of prostate gland
 9. Inferior ramus of pubic bone
10. Origin of adductor muscles
11. Spermatic cord
12. Femoral vein
13. Saphena magna vein
14. Femoral artery

15. Quadratus femoris muscle
16. Sciatic nerve
17. Inferior gluteal vessels
18. Tuberosity of ischium
19. Internal obturator muscle
20. Urethra (pars prostatica)
21. Levator ani muscle
22. Rectum
23. Ischiorectal fossa
24. Internal pudendal vessels
25. Prostate gland
26. Gluteus maximus muscle
27. Neck of femur

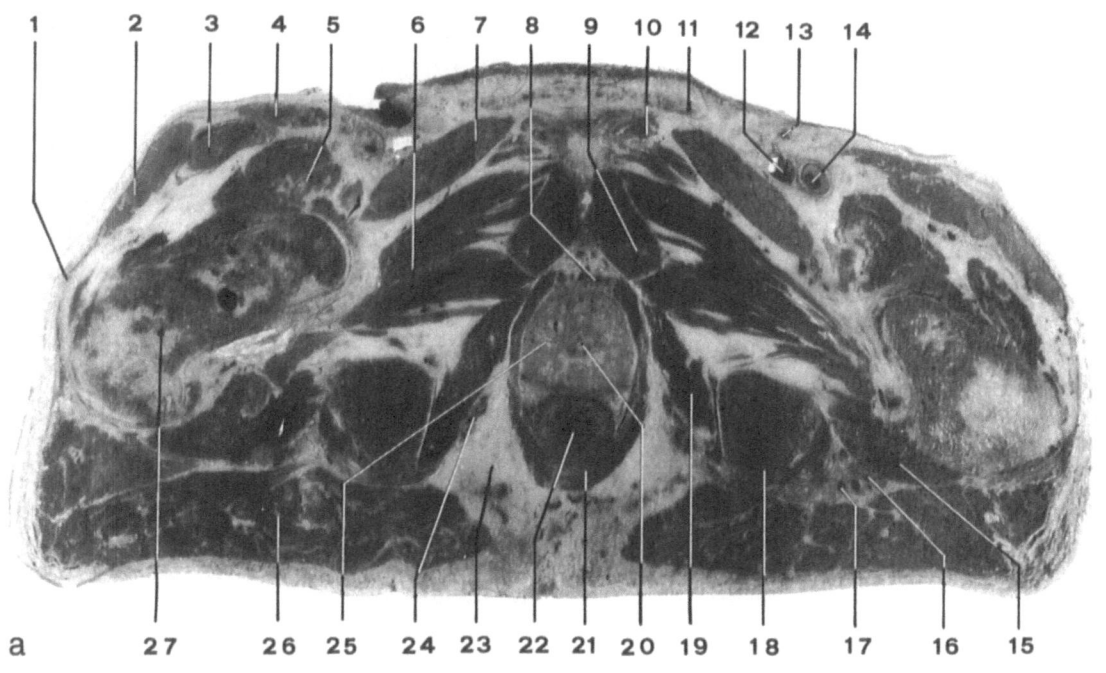

a

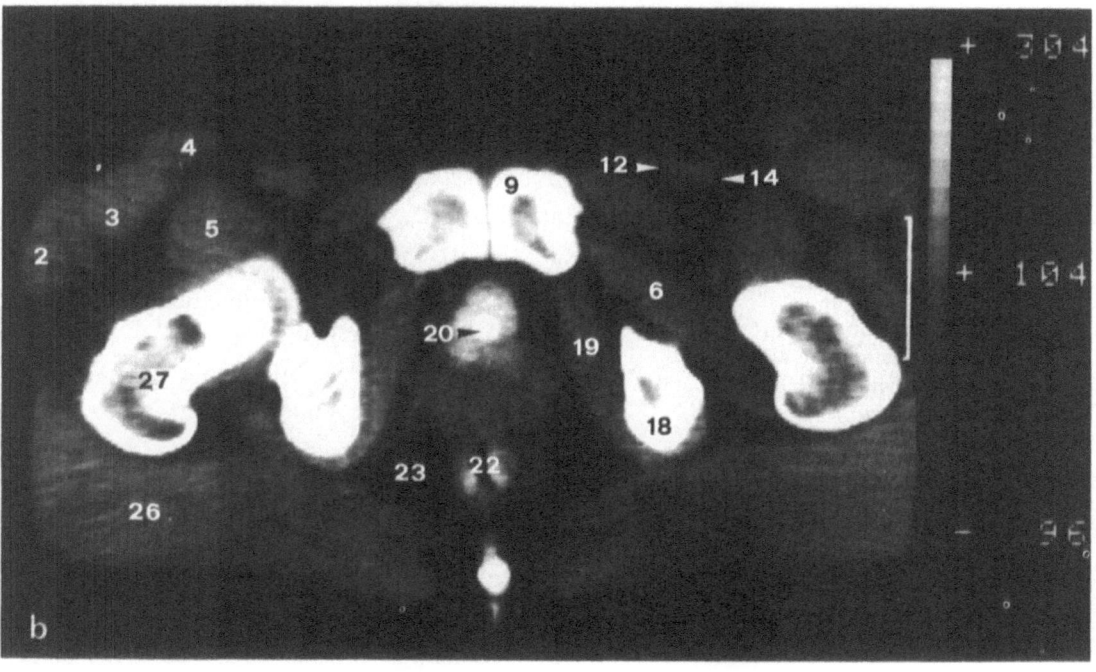

b

3.5.2 Female Pelvis

The structures of the pelvic wall and of the hip region are similar in both sexes (Fig. 44). Within the pelvic cavity and the levator funnel lie the bladder (ventral) and the rectum (dorsal). Between these structures lie the uterus and vagina. The broad ligament lies lateral to the uterus and extends across the pel-vic cavity in a frontal plane. The uterine tube is enclosed in the free edge of this ligament; the dorsal aspect of the ovary is attached to the broad ligament via the mesovarium. Usually the ovary lies in the ovarian fossa, ventral to the bifurcation of the common iliac artery.

Figure 44 shows the most commonly used planes of section in the clinical examination, and the transected pelvic organs at each level.

Fig. 44. Female pelvis. *a.* Frontal section. *b.* Lateral aspect.

A–C Planes of section
1. Abdominal aorta
2. Right common iliac artery
3. Right internal iliac artery
4. Right external iliac artery
5. Inferior mesenteric artery
6. Inferior vena cava
7. Right internal iliac vein
8. Right external iliac vein
9. Right ovarian vein
10. Left ureter
11. Iliacus muscle
12. Psoas major muscle
13. Internal obturator muscle
14. Levator ani muscle
15. Inferior ramus of pubic bone
16. Urogenital diaphragm
17. Vagina
18. Urethra
19. Urinary bladder
20. Uterus
21. Uterine tube
22. Fimbriae tubae
23. Ovary
24. Rectum
25. Ilium
26. Hip joint
27. Head of femur

1
2
3
4
5
6
7
8
9
10
11
12
13
14
15
16
17
18
19
20
21
22
23
24
25
26
27
A
B
C

a

9
10
2
3
4
22
20
23
24
19
17
18
A
B
C

b

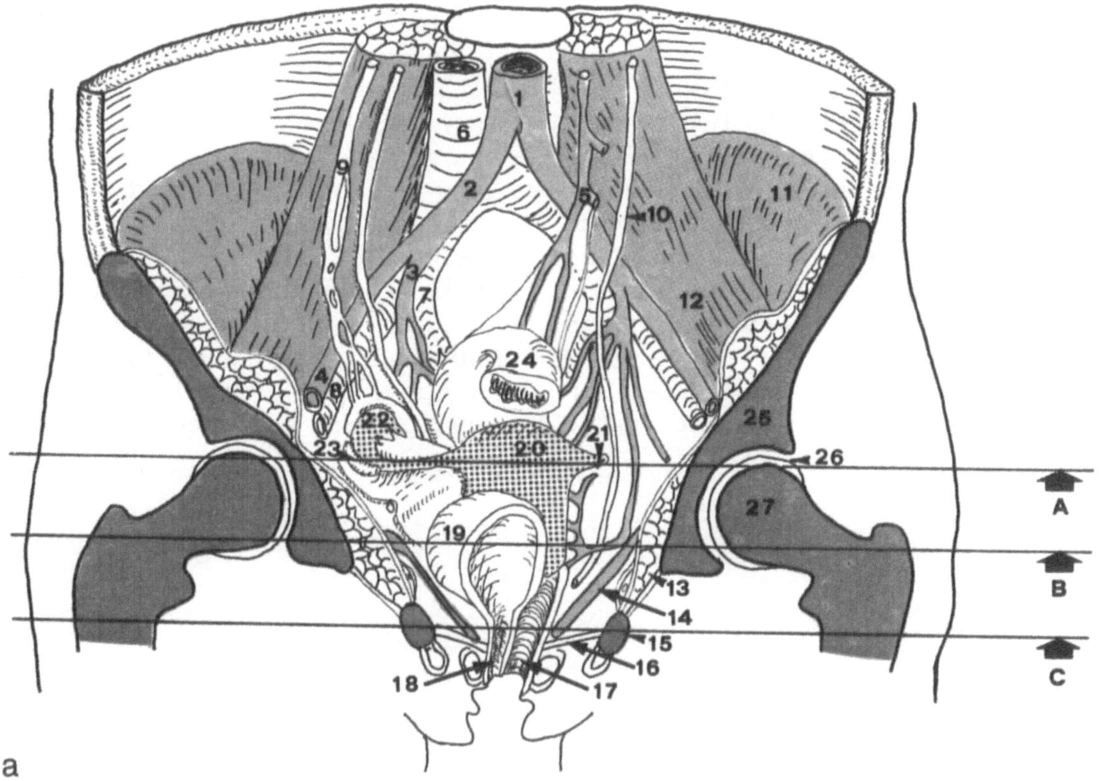

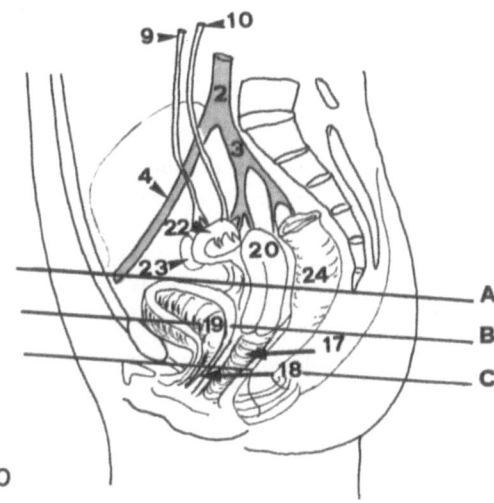

3.5.2.1 Horizontal Section A

The plane of section is through the fundus of the uterus, the two ovaries, and the hip joint (Fig. 45). The sigmoid colon, the fundus of the uterus, and the rectum are located in the center of the section. On the right the ovary and, dorsal to it, the fimbriae are seen adjacent to the medial surface of the ilium. The uterus is displaced to the left. The two uterine tubes, invested with mesosalpinx, are visible bilaterally.

A large loop of the sigmoid colon lies directly dorsal to the abdominal wall, another loop on the right of the uterus. Figure 45a depicts the course of the sigmoid colon.

The acetabula, sectioned femoral heads, and gluteal muscles are visible bilaterally.

In the CT image the bladder is filled with contrast medium, as are the external iliac lymph nodes (after lymphography); the body of the uterus, mesosalpinges, and rectum are visible.

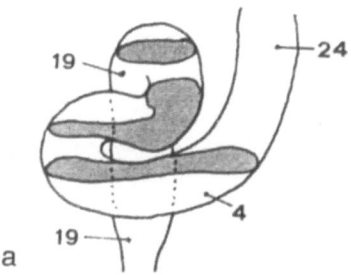

Fig. 45. Female pelvis. *a.* Course of sigmoid colon in this case. *b.* and *c.* Horizontal section A.

1.	Tensor fasciae latae muscle	15.	Gluteus minimus muscle
2.	Sartorius muscle	16.	Gluteus medius muscle
3.	Psoas muscle	17.	Sciatic nerve
4.	Sigmoid colon	18.	Inferior gluteal vessels
5.	Uterus	19.	Rectum
6.	Sigmoid colon	20.	Coccyx
7.	Convolution of ileum	21.	Os coxae
8.	Inferior epigastric vessels	22.	Head of femur
9.	External iliac vein	23.	Hip joint
10.	External iliac artery	24.	Descending colon
11.	Femoral nerve	25.	Urinary bladder
12.	Mesosalpinx	26.	External iliac lymph nodes
13.	Ovary	27.	Sacrum
14.	Gluteus maximus muscle		

1
2
3
4
5
6
7
8
9
10
11
12
13
23
22
21
13
20
19
18
17
16
15
14

b

25
26
26
21
5
19
27

c

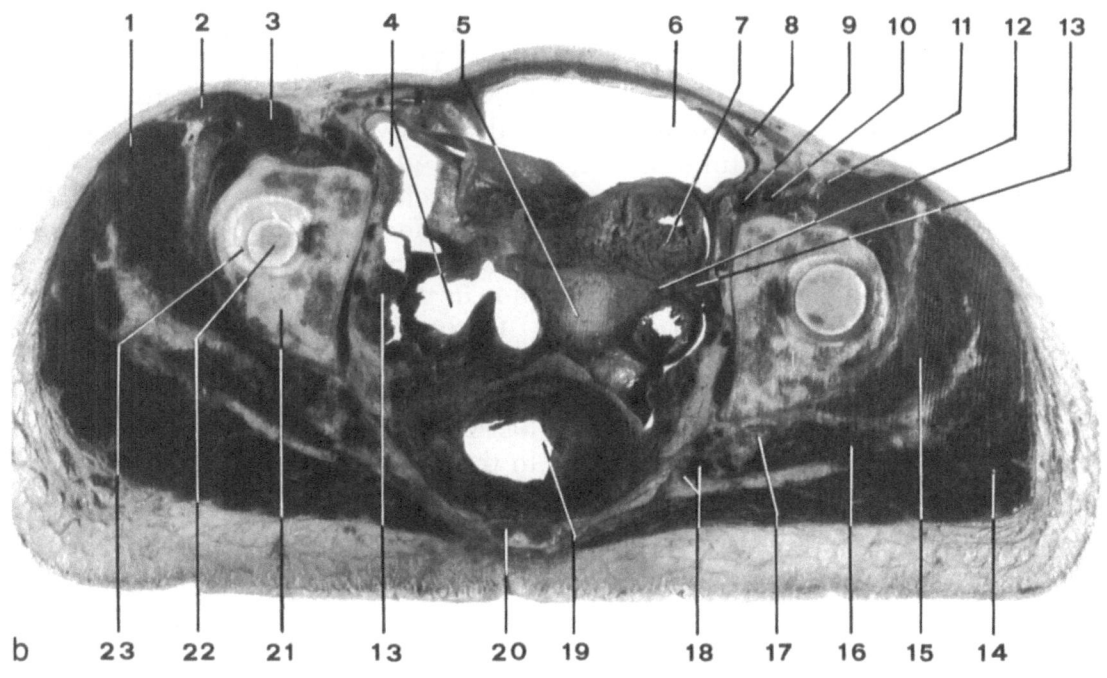

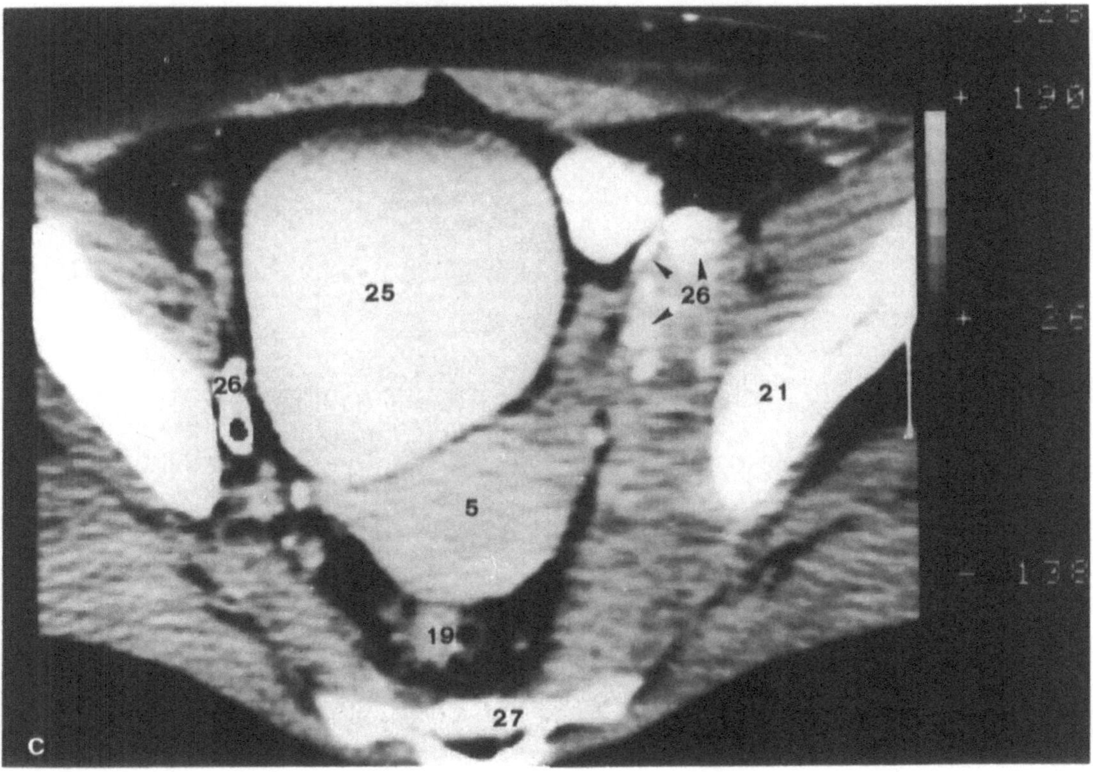

3.5.2.2 Horizontal Section B

The plane of section is through the femoral heads, the ischial tuberosities, the levator funnel containing the apex of the bladder, the cervix of the uterus, the fornix of the vagina, and dorsally the rectum (Fig. 46). A loop of sigmoid colon is transected on the right side of the bladder. The internal obturator muscle is seen on the medial surface of the os coxae, on the lateral surface is the hip joint. The internal obturator muscle can be followed around the is-chial tuberosity into the trochanteric fossa. The ischiorectal fossa lies between the internal obturator muscle and the levator ani muscle. Between the two gluteus maximus muscles the ischiorectal fat communicates with the subcutaneous fat. The femoral vessels lie ventromedial to the head of the femur in the iliopectineal fossa.

In the CT image the bladder is indented by the uterus, which is shifted to the right; the rectum, containing barium contrast medium, lies ventral to the coccyx. Laterally the hip joints are visible.

Fig. 46. *a* and *b*. Female pelvis: horizontal section *B*.

1.	Tensor fasciae latae muscle	15.	Neck of femur
2.	Hip joint	16.	Greater trochanter
3.	Sartorius muscle	17.	Lesser trochanter
4.	Iliopsoas muscle	18.	Sciatic nerve
5.	Femoral nerve	19.	Internal obturator muscle
6.	Obturator vessels	20.	Internal pudendal vessels
7.	Pectineus muscle	21.	Cervix of uterus
8.	Sigmoid colon	22.	Vagina
9.	Urinary bladder	23.	Rectum
10.	Superior pubic ligament	24.	Levator ani
11.	Femoral vein	25.	Ischiorectal fossa
12.	Femoral artery	26.	Ischial tuberosity
13.	Ligament of head of femur	27.	Gluteus maximus muscle
14.	Head of femur	28.	Uterus (CT section)

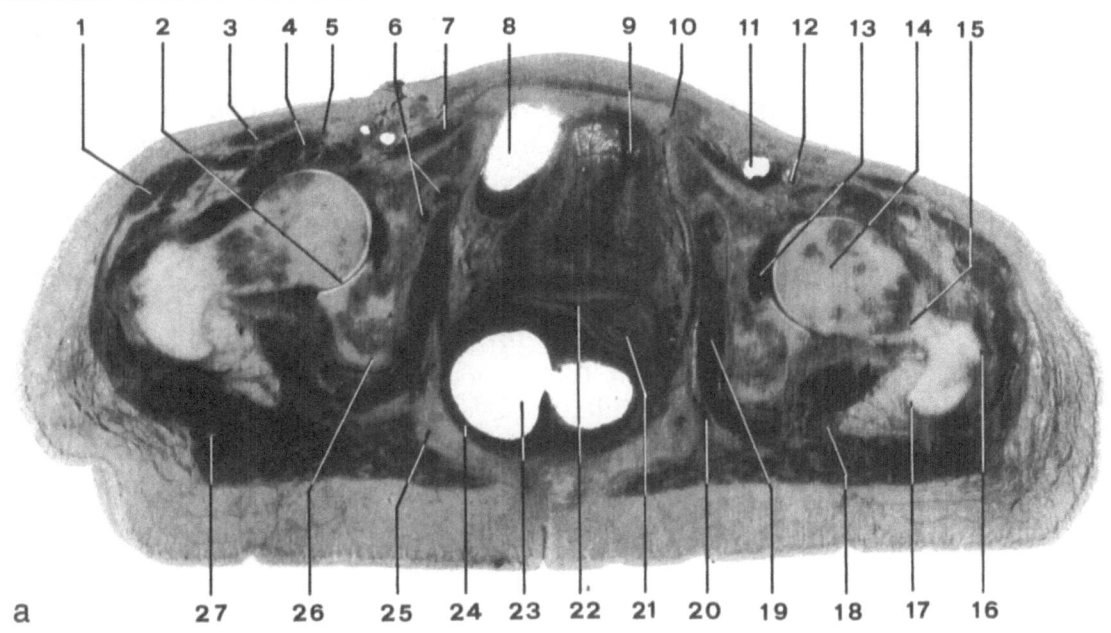

a

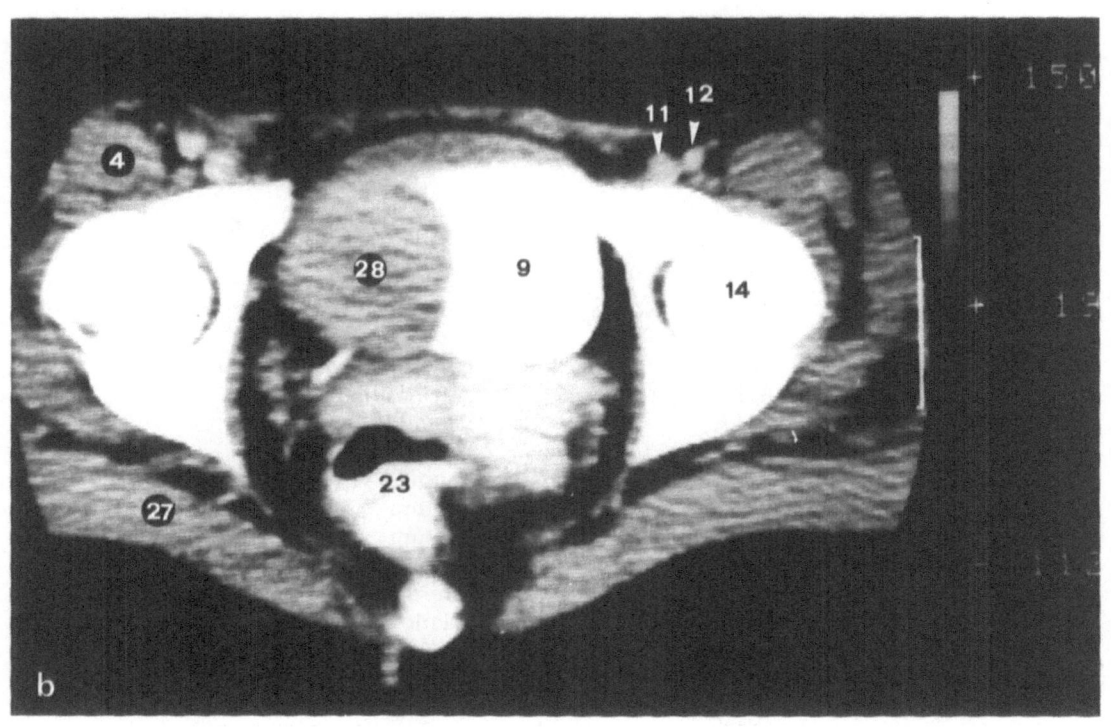

b

3.5.2.3 Horizontal Section C

The plane of section is through the symphysis, the neck and greater trochanter of the femur, the urethra, and the vagina (Fig. 47). Within the levator funnel lie the urethra, vagina, and rectum. The internal obturator muscle is seen between the inferior ramus of the pubic bone and the ischial tuberosity. The ischiorectal fossa, filled with fat, is seen between the levator ani muscle, internal obturator muscle, and gluteus maximus muscle. The external obturator muscle lies between the pectineus muscle and the internal obturator muscle. In the subinguinal region the following vessels are transected: great saphenous vein, femoral vein, profunda femoris artery, and femoral artery.

In the CT image, with the exception of vessels and nerves, most of the above listed structures are visible.

Fig. 47. *a* and *b*. Female pelvis: horizontal section *C*.

1.	Iliotibial tract of fascia lata	*14.*	Gluteus maximus muscle
2.	Tensor fasciae latae muscle	*15.*	Neck of femur
3.	Rectus femoris muscle	*16.*	Ischial tuberosity
4.	Sartorius muscle	*17.*	Ischiorectal fossa
5.	Iliopsoas muscle	*18.*	Vagina
6.	External obturator muscle	*19.*	Urethra
7.	Pectineus muscle	*20.*	Rectum
8.	Inferior ramus of pubic bone	*21.*	Levator ani muscle
9.	Great saphenous vein	*22.*	Internal obturator muscle
10.	Femoral vein	*23.*	Sciatic nerve
11.	Lateral femoral circumflex artery	*24.*	External obturator muscle (insertion)
12.	Profunda femoris artery		
13.	Greater trochanter		

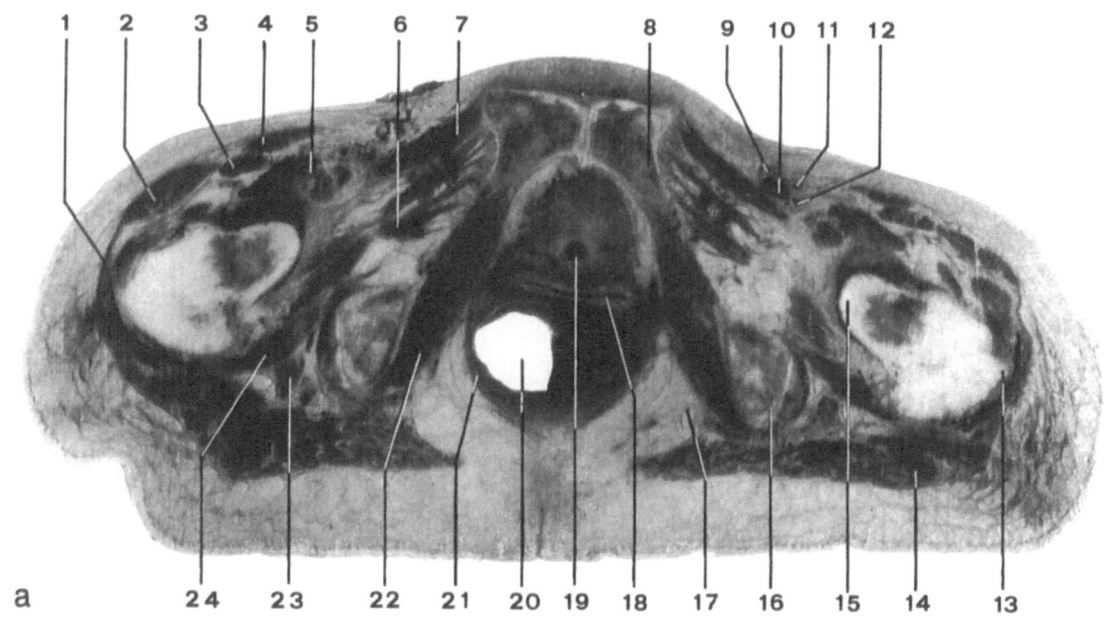

a

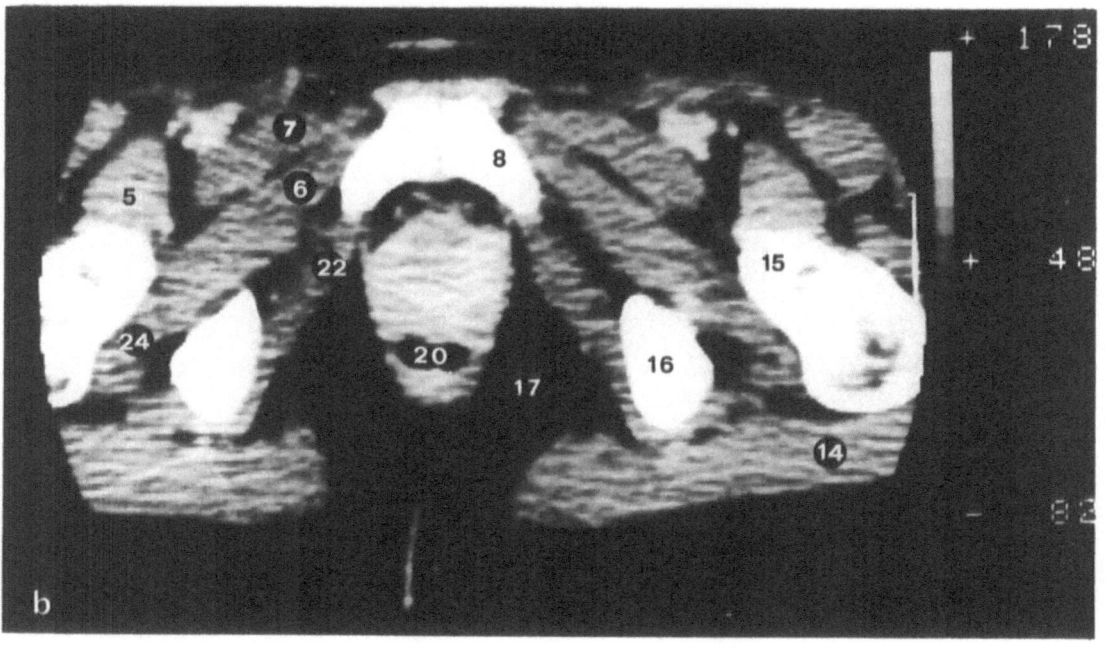

b

References

Bo WJ, Meschan I, Krueger WA: Basic Atlas of Cross-Sectional Anatomy. Saunders, Philadelphia, 1980

Chiu LC, Lipcamon JD, Yiu-Chiu VS: Clinical Computed Tomography: Illustrated Procedural Guide. Aspen Publishers, Rockville, MD, 1986

Fix JD: Atlas of the Human Brain and Spinal Cord. Aspen Publishers, Rockville, MD, 1987

Gambarelli J, Guerinel G, Chevrot L, Mattei M: Computerized Axial Tomography. Springer, New York, 1977

Haaga JR, Alfidi JR: Computed Tomography of the Whole Body, Vols I–II. Mosby, St. Louis, 1983

Han MC, Kim CW: Sectional Human Anatomy. Ilchokak, Seoul, Korea, 1985

Hollinshead WH: Anatomy for Surgeons, Vols I–III. Harper & Row, New York, 1966

Hounsfield GN: Computerized transverse axial scanning (tomography) Part I Description of system. Br. J. Radiol. 46: 1016–1022, 1973

International Anatomical Nomenclature Committee of the 11th International Congress of Anatomists: Nomina Anatomica, ed 5. Williams & Wilkins, Baltimore, 1983

Kretschmann H-J, Weinrich W: Neuroanatomy and Cranial Computed Tomography. Georg Thieme, Stuttgart, 1986

Lee SH, Rao Krishna CVG: Cranial Computed Tomography and Magnetic Resonance Imaging. McGraw-Hill, New York, 1986

Moore KL: Clinically Oriented Anatomy, ed 2. Williams & Wilkins, Baltimore, 1985

Morgan CE: Basic Principles of Computed Tomography. University Park Press, Baltimore, 1983

Romanes GJ: Cunningham's Textbook of Anatomy, ed 12. Oxford University Press, Oxford, 1981

Williams PL, Warwick R: Gray's Anatomy, ed 36 (British). Saunders, Philadelphia, 1980

Woodburne RT: Essentials of Human Anatomy, ed 7. Oxford University Press, New York, 1983

Index